JCER

Journal of Consciousness Exploration & Research

Volume 3 Issue 10

November 2012

Quantum Aspect of Psychiatry & Foundation of Reality

Editors:

Huping Hu, Ph.D., J.D.

Maoxin Wu, M.D., Ph.D.

Advisory Board

Dainis Zeps, Ph.D., Senior Researcher, Inst. of Math. & Computer Sci., Univ. of Latvia, Latvia
Matti Pitkanen, Ph.D., Independent Researcher, Finland
Arkadiusz Jadczyk, Professor (guest), Center CAIROS, IMT, Univ. Paul Sabatier, Toulouse, France
Alan J Oliver, Independent Researcher, Australia
Sultan Tarlacı, M.D., Neurology Specialist, NeuroQuantologist, Turkey
Gregory M. Nixon, University of Northern British Columbia, Canada
Stephen P. Smith, Ph.D., Visiting Scientist, Physics Dept., UC Davis, United States
Elio Conte, Professor, Dept. of Neurological and Psychiatric Sciences, Univ. of Bari, Italy
Michael A. Persinger, Professor, Laurentian University, Canada
Andrei Khrennikov, Professor, In'tl Center for Mathematical Modeling, Linnaeus Univ., Sweden
Chris King, Independent Researcher, New Zealand
Graham P. Smetham, Independent Researcher, United Kingdom
Steven E. Kaufman, Independent Researcher, United States
Christopher Holvenstot, Independent Researcher in Consciousness Studies, United States
Pradeep B. Deshpande, Prof. Emeritus of Chemical Engineering, Univ. of Louisville, United States
Iona Miller, Independent Researcher, United States

ISSN: 2153-8212   Journal of Consciousness Exploration & Research   www.JCER.com
Published by QuantumDream, Inc.

Table of Contents

Articles

Possible Roles of Cell Membrane & Cytoskeleton in Quantum Aspect of Psychiatry
Massimo Cocchi, Lucio Tonello & Fabio Gabrielli   01-19

Foundation of Reality: Total Simultaneity
Wilhelmus de Wilde   20-28

A Metaphysical Concept of Consciousness
Wilhelmus de Wilde   29-42

Explorations

Self-Awareness and Memory
Narendra Katkar   43-51

Consciousness Dimensions & Quantum Non-locality
Cebrail H. Oktar   52-63

Article

Possible Roles of Cell Membrane & Cytoskeleton in Quantum Aspect of Psychiatry

[1,2]Massimo Cocchi*, [1]Lucio Tonello & [1]Fabio Gabrielli

[1]"Paolo Sotgiu" Institute for Quantitative & Quantum Psychiatry & Cardiology
L.U.de.S. Univ., Lugano, Switzerland
[2]Dept. of Medical Veterinary Sciences, University of Bologna

Abstract

This paper is concerned with the molecular aspects of the classification of subjects with Major Depression (MD) and Bipolar Disorder (BD) and the possible roles of cell membrane and cytoskeleton in quantum aspect of psychiatry. Classification is based on a Self Organizing Map (SOM), called ADAM, of three fatty acids in platelets (Palmitic acid-PA [C16:0], Linoleic Acid-LA [C18:2], Arachidonic Acid-AA [C20:4]) and their roles as markers of the two major mood disorders (MD and BD). The weighted considerations on the dynamics between membrane mobility and interactiome and the symmetry breaking between MD and BD seem to be linked to Linoleic Acid. The latter is critical in the regulation of molecular fine tuning apparently involved in modulating different levels of intensity of mood disorders, consciousness, learning and memory processes.

Keywords: Platelets, Fatty Acids, Linoleic Acid, Interactome, Symmetry Breaking, Major Depression, Bipolar Disorder, Consciousness.

Introduction

Brain cell membrane seems to offer a promising path toward a better understanding of psychopathology in its diferrent aspects: Major Depression, Bipolar Disorder and their consequences on cognitive impairment (Austin et al. 1992; Bearden et al. 2006; Burt et al. 1995, Deckersbach et al. 2004; Basso and Bornstein 1999; William et al. 1997; Breslow et al. 1980; Colby and Gotlib 19887; Golinkoff and, Sweeney 1989; Stromgren 1977). Diverse approaches have provided evidence with respect to this topic. In particular, a wide scientific literature supports the suggestion of Rasenick's group (Allen et al. 2007; Donati et al. 2008;) that human platelets serve as biological markers for depression (Mooney et al. 1988; Mooney et al. 1998; Garcia-Sevilla et al. 1990; Pandey et al. 1990; Menninger and Tabakoff 1997; Hoffman et al. 2002; Hines and Tabakoff 2005).

Other authors, namely Cocchi and Tonello, have studied platelet membranes of depressed subjects, enlisting profiles of Fatty Acids (FAs) as a possible measure of the membrane status (Cocchi and Tonello 2006a; Cocchi and Tonello 2007a; Cocchi and Tonello 2007b; Cocchi et al. 2008; Cocchi et al. 2009a; Cocchi et al. 2009b; Cocchi et al. 2006b). In particular, they have compared the platelet membrane FA profile of a group of clinically depressed subjects against a group of healthy subjects. The pattern of the two groups has been found to be statistically different. The FA findings in the authors' experiments appear highly consistent with the G-protein and lipid raft dynamics delineated by Rasenick's group (Chen et al. 2007; Sottocornola and Berra 2008;

---

* Correspondence: Professor Massimo Cocchi, Dept. of Medical Veterinary Sciences, Univ. of Bologna, Via Tolara di Sopra 50, 40064 Ozzano dell'Emilia, Bologna. E-mail: massimo.cocchi@unibo.it

Stulnig et al. 2001). Recent growth of clues and evidence points toward a link between FAs and G-protein dynamics (Han et al. 2009; Bok et al. 2009). The authors of the present paper have performed further analysis: data studied by means of a Self Organizing Map (SOM)-the so called "Kohonen Network" - (Kohonen 2001), suggest three FAs as the main actors in depression (and suicide) and thus as possible unique platelet biomarkers. They are arachidonic, palmitic and Linoleic Acid. The authors hypothesize that these three FAs might suffice to summarize the whole cell membrane status, i.e. the full FA profile, and hence the G-protein dynamics. So, the cell membrane seems to be deeply involved in depression and perhaps also in other neuropsychiatric illnesses. In particular, G-protein dynamics in brain cells and FA profiles in platelet membranes could be a very promising way to better understand psychopathology.

These two molecular entities seem to give different measures of the same thing, the brain cell locus being of central importance but the platelet site perhaps easier practical access. It has been recently shown (Tonello and Cocchi 2010) that there are many clues suggesting multiple strong connections between FA profiles or G-protein dynamics and current quantum brain theories. So, cell membrane can be seen as important building block for the most notable quantum models of brain, mind and consciousness. In particular, FA profiles or G-protein dynamics seem to be a common foundational element shared by all of these quantum models and theories.

The quantum world may be useful in better understanding psychiatric illness, but psychiatry as well may contribute to improving the quantum research applicable to consciousness. In both cases, the cell membrane might be a good starting point, linking the two worlds in a powerful way.

A long path of bio-molecular and mathematics research, has been successfully completed, in an attempt to coordinate and interpret the "unusual" analytical results obtained from the platelets fatty acids analyses, which are representative of the composition of the membrane. The term "unusual" means to emphasize the fact that, today, the studies on fatty acids are a bit obsolete in view of the international scientific research. Perhaps, too soon, has been forgotten that they play in phospholipid bilayer together with cholesterol, a key role in the complex mechanism of regulation of the membrane mobility. The rigorous interpretation of the results obtained has stimulated awareness that it was possible to link the experimental findings to membrane quantum aspects of the psychopathological phenomenology. Below, will be addressed the main steps and findings of the research to demonstrate the existence of a symmetry breaking between MD and BD. Evidence, the latter, which called for a strong reasoning on the role of Linoleic Acid.

Considerations on the results achieved in humans

If one considers the strong power of the SOM classification, the platelets fatty acids which have been shown to have discriminating power between Major Depression (MD) and Bipolar Disorder (BD), and that through the calculation of an index (B2) that relates the percentages of fatty acids with their melting point and molecular weight, it was possible to recognize the MD from the BD, seems a foregone conclusion that the SOM and the index calculated (B2) read a biological phenomenon among the most complex, the mood disorder, with predictive capabilities, properties and diagnostic accuracy.

In the scientific literature produced (Cocchi et al. 2008; Cocchi and Tonello 2010a; Cocchi and Tonello 2012) were discussed in great detail the molecular aspects of mood disorders (Cocchi and Tonello

2010b), read and interpreted in the dynamics of the interactome, i.d. in the snapshot that is sharing the mobility of the membrane, protein Gsa and cytoskeleton (Cocchi et al. 2010c).
On this basis was discussed and formulated the hypothesis of the link among interactome, classic and quantum consciousness also from the philosophical point of view (Cocchi et al. 2011a; Cocchi et al. 2011b).

Nevertheless, there are still gray areas, not on the certainty of the phenomenon, but on its inner meaning. What follows is an attempt to clarify the path of the results obtained. The first step is to observe in detail the SOM and the figure that has distributed the B2 index. We will do so with the knowledge that it was not and it is not possible to manipulate the results in the light of the inviolability which is characteristic of the mathematical function that builds the SOM.

As Donald Mender, a psychiatrist in Yale, has observed, "the numerical data may represent a new and empirical truth", aware that the SOM and the B2 framework have nothing in common between them, by the mathematical and biological point of view. Nevertheless, it is not possible to classify a human or animal subject in its clinical diagnostic correspondence, without the use of both tools, since the SOM is two-dimensional and the B2 map one-dimensional, being that the SOM considers three numeric entities and the B2 map considers numeric entities obtained from parameters that the SOM could not know, such as the molecular weight and the melting point of each fatty acid.

Let us begin with the analysis of the SOM (Figure 1) as it appeared the first time, trying to explain the distribution of cases read simultaneously with the index B2 (Figure 2).

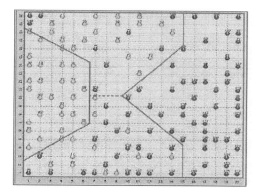 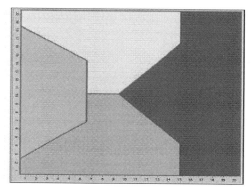

Figure 1. Representation of SOM and subdivision of the main areas of classification of subjects studied.
Areas: Green = Normal, Red = Pathological, Yellow = high density of normal, Orange = high density of pathologic

The observation, from which will emerge most of the considerations, was linked to the different levels and gradualness of saturation/unsaturation ratio of platelet fatty acids. The fatty acids identified by the SOM were PA (saturated), LA and AA (polyunsaturated), their sum was constant, for all cases, and greater than 50% (53.33 ± 3.43%), representing the majority of the whole fatty acids profile. A different degree of mobility of the platelet membrane, therefore, appeared to be a characteristic finding within mood disorders (MD and BD) and between them and the condition of apparent normality.

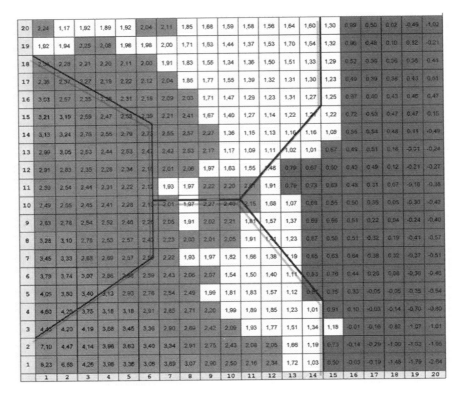

Figure 2. B2 index distribution of a one-dimensional map with overlapping areas identified by the SOM (two-dimensional) for the classification of subjects studied.

An accurate diagnostic investigation, together with a considerable increase in cases, confirms that the negative index corresponds to cases of MD and the positive to cases of BD (Benedetti et al. 2012). This result clarifies the overall picture of the distribution of cases and allows hypotheses and conjectures on the intensity of mood disorders.

The reconstruction of the exchange of arachidonic acid between platelet and brain in pigs (Cocchi et al. 2009a; Cocchi et al. 2009b), enables the understanding of a key point:
The distribution of cases obtained on the SOM and B2 map, consistent with higher or lower levels of mobility of the platelet membrane, explains how the serotonin uptake by cell (neuron/platelet) is modulated by such chemical characteristics, ie, lower mobility higher uptake and vice versa (Benedetti et al. 2012).

The fact, however, that neuron and platelet (both derived from ectoderm), are two entities completely separate and autonomous with different functions, together with the fact that serotonin does not cross the blood-brain barrier, makes it plausible the demonstration that is the transfer of AA between the two cells, the critical point which conditions their membranes mobility, and then the modulation of the passage of serotonin. We will explain, now, why it is believed that Linoleic Acid plays a decisive role about the placement of the subjects on the SOM and B2 index map.

Linoleic acid: molecular fine tuning regulator of mood disorders?

Why you can read in platelets, what happens in neurons, in the case of mood disorders, still remains a complex issue with respect to which it does not seem possible to give a comprehensive answer. The configuration of the level curves of the various fatty acids realized in the SOM has allowed us to understand something of the phenomenon (Figure 3). One glimmer of light appears when we realize that only with the SOM and B2 map, together, we can accurately identify the characteristics of the subject (SOM and B2 do not know of each other but converge on the same target).

This observation is necessary when for more than one subject, with the same index, we will have two different positions in the SOM, i.d., the same index can classify subjects in different points of the SOM. This finding is very important because it means that, in addition to the reasoning on the mobility of the membrane, there is a different element which condition the result, and since everything revolves around the three fatty acids mentioned above, must necessarily be one of them, in its concentration, that makes the difference and can influence the condition of the subject, being equal the mobility of the membrane. Even in these cases we have accurate diagnostic findings.

For a number of reasons that we will try to make explicable, a particular attention is drawn to the Linoleic Acid, an essential fatty acid, which could not be produced by the organism, whether human or animal. The level curves, show that the absolute minimum of Linoleic Acid is at the point shown below (yellow arrow) and which corresponds, also, to the minimum point of the B2 index (Figure 3):

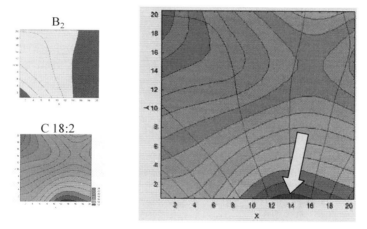

Figure 3. B2 Index and Linoleic Acid distribution on the SOM

The fan produced by the SOM from right to left shows that the B2 is in progression from -2.64 to 8.23 (see Figure 3). The healthy subjects belonging to the green area are characterized by a mean value of B2 equal to 2.80. This value is the midpoint between the extremes -2.64 (absolute minimum expressed by the map) and 8.23 (absolute maximum expressed by the map). A careful analysis of the mathematical formulation of the B2 index shows that it is governed almost entirely by the Arachidonic Acid and Palmitic Acid. Intuitively, this can be deduced by expressing the mean values of the two fatty acids on the SOM using the level curves which appear qualitatively symmetrical (Figures 4 and 5).

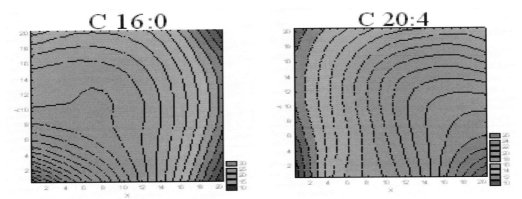

Figures 4 and 5. Demonstration of the symmetric distribution of Palmitic Acid (C16: 0) and Arachidonic Acid (C20: 4).

These two fatty acids determine the macro areas which carachterize the subject position. Within a macro area is the Linoleic Acid which modulates the precise position of the subjects. Is C18: 2 the main actor in the "fine tuning"? (Figure 6)

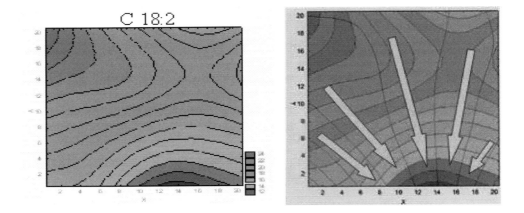

Figure 6. Distribution of Linoleic Acid on the map. The blue position represents the core of the SOM

A practical example: two subjects with the same B2 index, one diagnosed with BD (located in the orange area with a B2 positive index) and the other normal (located in the green area) are characterized by a significant difference in the proportion of Linoleic Acid.

It should be noted the plausibility of this observation if one considers the low value of the melting coefficient of Linoleic Acid. It is likely that in the circumstances that discriminate a healthy from a pathological subject, the difference of Linoleic Acid determines, however, a modification of the biochemical and molecular factors involved and/or responsible for the biochemical and pathological determinism of the mood disorders. It should be noted, also, that the effects of Linoleic Acid on the pathology are relevant for configurations that see equally balanced Arachidonic and Palmitic Acid.

Practically, with reference to the positions of the subjects in the SOM, for values beyond a certain limit of Arachidonic Acid the subject can be considered depressed, the same for values beyond a certain limit of Palmitic Acid. When B2 is in a neighborhood of the normal value, the percentage of Linoleic Acid becomes decisive.

These configurations introduce, however, the concept of hyper saturation of platelet in the case of an excess of Palmitic Acid and that of hyper unsaturation in the case of an excess of Arachidonic Acid. A further discriminant, between the two fatty acids mentioned, for the recognition of the pathology, must be identified in the Linoleic Acid which, in case of excess, redetermines the biochemical conditions for the development of the disease. Of course, the network operates beyond these operations, not yet known and probabilistically interpreted, of connection among the values that were administered.

Some considerations on brain fatty acids (AP, AL, AA) with respect to the SOM

With respect to platelet, the mathematical function (SOM) has read the values of all fatty acids and interpreted them without knowing their biological significance. The knowledge that even in the human brain the Linoleic Acid percent is very low (**Svennerholm 1968; Kei et al. 2012**, as well as in the rat (**Bourre et al. 1990; Barzanti et al. 1994**), pig (**Cocchi et al. 2009a, b**) and in chick embryo (**Maldjian et al. 1995; Noble and Cocchi 1990**) has urged the attempt to see how the SOM would have positioned the three platelet fatty acids (PA, LA, AA) obtained by pig brain, just as was for the platelets (Figure 7).

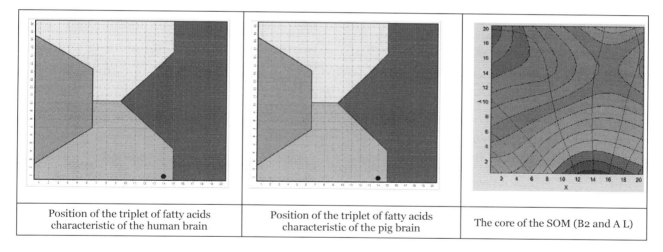

| Position of the triplet of fatty acids characteristic of the human brain | Position of the triplet of fatty acids characteristic of the pig brain | The core of the SOM (B2 and A L) |

Figure 7. Position of the brain of man and pig with respect to the SOM

There is no doubt that Linoleic Acid is the element which conditions the distribution of the weights between AA and PA as well as the representation of the fatty acids triplet in the brain corresponds to the core of the SOM (Figure 8).

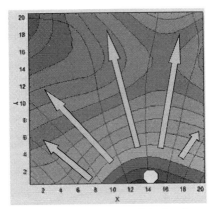

Figure 8. The fatty acids triplet (PA, LA, and AA) of the brain occupies the position of the SOM core

The level curves also show how, unlike other fatty acids, only Linoleic Acid is distributed in the whole SOM, ie, with a calibrated progression of levels, as if it was the regulator of the fine tuning of B2 index with all the biological and chemical implications that it involves.

With respect to what emerges from the above considerations opens a series of reflections is opened:

1. The brain appears to be structurally predisposed to be pathologic, but the pathology needs to be expressed, and this is partly due to the balance of Palmitic and Arachidonic Acid which should maintain a normal condition of the membrane mobility, through the modulation of Linoleic Acid. It has been demonstrated that in depression, Arachidonic Acid is high in the brain (Green et al. 2005).
2. The Linoleic Acid, in the brain, has a very low concentration and appears to be the stabilizing element offsetting the balance between Palmitic and Arachidonic Acid, which, otherwise, could favor dangerous modifications of mobility.
3. From the metabolic point of view, it would seem to be a defect of the desaturase activity (predominantly Delta 6/Delta 5) in the sense of an excess, in pathological subjects, which favors a substantial formation of arachidonic acid, which in turn is conveyed to the neuron. The desaturases, therefore, in the case of mood disorders would behave the opposite of what is expected to the passing of the age.
4. One can not, however, exclude the possibility that in addition to this aspect there is a strict control of the mechanisms of incorporation of Linoleic Acid at the cellular level, in certain conditions.

Certainly Linoleic Acid is a critical point, and in this sense we can even take help from the animal world. The experiments conducted on different animals and the evaluation of the data, using the SOM, as well as for humans, has provided evidence of extraordinary correspondence between man and animal, with regard to mood disorders (Cocchi et al. 2009c; Cocchi et al. 2012).

Linoleic Acid is significantly lower in animals that are placed in the area representative of the pathology and relatively high in animals that occupy the opposite side (Figure 9):

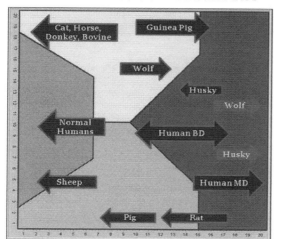
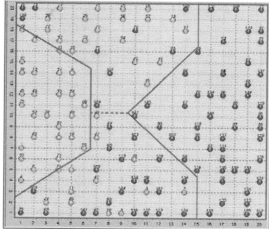

Figure 9. Distribution of the animals on the Depression Map (SOM) in comparison with the original human depression map. The dark blue arrow represents the position of the average values of the B2 index of the animals concerned and humans. The light blue arrow shows the position of a specimen of a German shepherd and Alaskan Malamunte in the area with a negative index, then, corresponding to that area which placed human subjects with major depression. In this way, it was observed that some animals (sheep, cat, horse, donkey, cattle) range in opposite area to the one identified by the SOM for human pathological cases, while the rat, guinea pig and pig are at areas identified by the SOM for pathological subjects.

Always remembering that Linoleic Acid is "essential" and when it is high, there may be only two interpretations:

- Large quantities should be introduced
- Desaturase activity is very low

It seems more plausible the second justification, also in the light of the disposition of the cat on the SOM, that, being a feline, by default, is shown to possess almost nothing delta 6 desaturase activity. Beyond all conjecture, Linoleic Acid, for direct or indirect reasons, is located at the center of phenomena that accurately influence its concentrations, both in humans and animals. A key point, the Linoleic Acid, which requires the opening of a new chapter of considerations. To better understand the reasoning on the data of Linoleic Acid, we must draw attention to the concept of symmetry breaking and some works that have linked an excess of Linoleic Acid, even at concentrations slightly higher, with some biological functions **(Marei et al. 2010)** and molecular interactions **(Namikoshi et al. 2002)**.

Further, with regard to a set of studies about cellular nutrition **(Cocchi et al. 1979)**, the effect of different amounts of phospholipids, extracted from various organs of calf (diencephalon, retina, cerebral cortex and heart), on chick embryo myocardium cell cultures were tested. From the numerous cell cultures it was observed that phospholipids, particularly at the highest levels, constantly decreased the culture migration velocity and this reduction was remarkably more accentuated in presence of heart phospholipids. Moreover in this condition the cells showed a change in their morphology and were full of large lipid drops, perhaps in consequence of an alteration of the chemical-physical plasmalemma characteristics and a change in the membrane enzyme activities involving lipid metabolism too. A far lesser effect was observed in cultures treated with phospholipids of the brain and retina. We did not realize at that time that the

cardiac phospholipids, unlike the others (diencephalon, retina and cerebral cortex), are very rich in Linoleic Acid, and that the addition of large amounts, to those naturally occurring, could be responsible for the deep changes seen. This effect, again, could confirm the criticality of Linoleic Acid.
Obviously, the consequences of the condition of excess will be relative to the biological system in which the phenomenon occurs.

At this point it seems appropriate to address the concept of symmetry breaking. We start from the definition of symmetry breaking given by the Nobel Prize P. W. Anderson: "Increased levels of symmetry breaking in many body systems (systems of many interacting elements) correlates with increasing complexity and functional specialization." Subsequently, the theory of symmetry breaking lead to an understanding of many phenomena both at the cellular and sub-cellular level **(Li and Bowerman 2010)**.

Recently it has been shown that the concept of symmetry breaking can describe the conditions that make the difference between Major Depression and Bipolar Disorder **(Cocchi et al. NeuroQuantology, in press)**. Perhaps within the concept of symmetry breaking and the considerations on the Linoleic Acid we can find answers to questions that have long affected the work. In particular, for a long time, the problem of how the set of the three fatty acids could correspond to a condition of MD or BD has been posed. The perception that the game of the molecular mechanisms identified could underlie implications of the quantum consciousness has been widely debated, finding aspects of great consistency in the molecular interactions involving membrane, protein Gsa and cytoskeleton **(Tonello and Cocchi 2010; Cocchi et al. 2010a; Cocchi et al. 2010c; Donati etal. 2008; Rasenick et al. 2004; Popova et al. 2002; Allenet al. 2007)**.

It has been tried, again, to demonstrate that those molecular mechanisms expressing biological responses of scalar complexity are of the same consistency in humans and animals: the complexity of the brain is expressed with different mode of language, but the molecular phenomenon remains unchanged. We have understood that the positive and negative sign of B2 index, which characterize respectively BD and MD, can never be interchangeable because of the demonstration of the symmetry breaking between MD and BD. For the data collected on Linoleic Acid and for their biological significance, it could be likely that is the Linoleic Acid to signify the concept of symmetry breaking.

Some scientific papers, albeit for other biological aspects, when Linoleic Acid is in excess, confirm its involvement with microtubules **(Namikoshi et al. 2002)** and show, in the event of excess, that it inhibits the growth mechanisms of oocyte and embryo **(Marei et al., 2010)**. The problem of the magnitude of oscillations of Linoleic Acid compared to the change in the levels of symmetry breaking arises, at this point, at least in the case of MD and BD. In the first remarks on the work we considered as correctly, to detect, how, in addition to the arachidonic and palmitic acid, a mood disorder could be expressed both for high and reduced levels of Linoleic Acid and this is confirmed in both human and animal data.

It would seem, at this point, possible, to understand how the SOM has analyzed the numbers. Evidently among all fatty acids available, the reasoning of the mathematic function that build the SOM is developed in relation to the critical distance of numbers that represent the percentages of Linoleic Acid, compared to those of the palmitic and arachidonic acid. Apparently none of the other remaining fatty acids, which, however, the SOM should have considered, showed such criticality.

Possible Roles of Cell Membrane & Cytoskeleton in Quantum Aspect of Psychiatry

Each cell membrane, within the phospholipid bilayer and according to the fatty acids profile, can be thought of as a system in constant dynamic state.

Each membrane, however, shall use all its tools in order to maintain the maximum stability, to avoid the risk of conditions that may be incompatible with cell life. From the point of view of the lipid composition the cell will implement all the resources to ensure the maintenance of the balance between cholesterol and phospholipids (50/50) and, within the fatty acid composition of the phospholipid fractions, the right compensation between saturated and unsaturated fatty

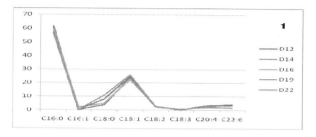
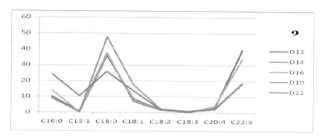
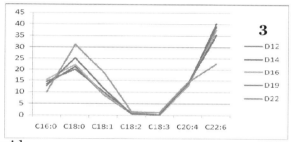
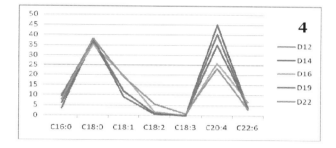

acids

Figure 10. Phospholipid fatty acids distribution during chick embryo brain development. It can be seen that even during the phases of development (D) the trend of fatty acids tends to maintain a substantial stability. Phosphatidyl – Choline (1), Phosphatidyl – Serine (2), Phosphatidyl - Ethanolamine (3), Phosphatidyl – Inositol (4).

A characteristic of the platelet is that, not having elongase and desaturase activity, must necessarily resorting to mechanisms of selectivity in the incorporation of fatty acids, in order to ensure the stability of the membrane. All this has been extensively investigated by observing the distribution of fatty acids in the platelets of hundreds of human and animal subjects and not entrusting the observation to the simple statistical evaluation, but using the SOM to evaluate the phenomenon of stabilization or no stabilization of fatty acids.

Considerations on platelets and neurons cell membrane because of the similarities found and extensively described in the literature as a bridge to the quantum consciousness, has resulted in subsequent reflections, due in part to the need to give plausibility to the theories and assumptions of the quantum consciousness *(Pioneers like the physicists Hiroomi Umezawa, Kunio Yasue, and Giuseppe Vitiello, mathematicians like Roger Penrose, and biomedical investigators like Stuart Hameroff, Gordon Globus, and Gustav Bernroider have plumbed the depths of subatomic structure and its macroscopic amplifications in search of substrates for quantum computation and other capabilities that may match attributes of the human psyche better than models advocated by conventional cognitive neuroscience, from QPP letter, 2012)* starting, as argued by Peter Bruza, from the concreteness of experimental data (Bruza 2010).

This concreteness, after a long period of research, was achieved when there was experimental evidence to recognize MD and BD through the membrane and its characterization in fatty acids. Research allowed to realize that the disappearance of the ability to maintain a stable membrane, in the case of mood disorders, may be cause of MD and BD. Disorders, these, which relate to the complex neuronal phenomenology and which involve aspects of consciousness. The cell membrane lives, therefore, in a constant condition of mobility and, when the oscillations of fatty acids cross the ranges of the membrane tolerance, pathological phenomena can be expressed.

The condition of constant mobility of the membrane, both, in physiological or pathological conditions, results in deformation of the same, exerting a mechanical force, at different levels of intensity with consequences that can involve membrane lipid raft microdomain, Gsα protein and cytoskeleton (Cocchi et al. 2010b). The dynamical factors of membrane lipids constitute lipid dependent factors that can dramatically affect protein functions or protein-protein interactions (McIntosh 2006). Phenomena of adhesion, between membrane and cytoskeleton, are influenced by a number of factors, including the lipid composition, cytoskeleton density and distribution, ratio between membrane surface area and volume, and internal cellular pressure. (Sheetz 2006).

In the literature, have been extensively described the cytoskeletal changes by forces exerted on the membrane (Chifflet and Hernandez 2011; Sun et al. 2007; Yap and Kamm 2005; Hoover et al. 1981; Apodaca 2002; Namikoshi et al. 2002; Janmey 1995; Janmey and McCulloch 2007). The study of quantum hypothesis of consciousness, the opportunity for dialogue that is open in the QPP, have certainly contributed to the construction of a new hypothesis attempt that could reconcile the previous ones which, in fact, are only the result of different ways of looking at the same phenomenon.

The state of continual renewal and exchange of fatty acids by the membrane with consequent deformation of the lipid bilayer, can be thought in terms of a mechanical force, in both physiological and pathological conditions, which is exercised on the cytoskeleton with continuity, allowing, when there are physiological conditions for the membrane, periods of decoherence of microtubules (500ms), which give rise to the conscious manifestation which will result in classical consciousness (neurocorrelate). The problem that arises is to understand how, under different conditions of mood disorders, the dechoerence can be modified by lengthening or shortening the physiologic decherence period. Having had the opportunity to understand that a higher and lower mobility of membrane, respectively, characterize MD and BD, it is perceived that there are pathological conditions of deformation of the membrane that may exert different mechanical forces on the cytoskeleton from the inside of the same membrane and could potentially affect the physiological phenomenon of superposition of microtubules and consequently the period of oscillation. The hypothesis is that in the case of MD and BD there would be, respectively, an elongation and a reduction of the period of decoherence. Modifications of the consciousness state would correspond to this possibility? This force, which in the case of the neuron could be called "neuro mechanical" could represent in fact, the conditioning factor of the microtubules activity and of their critical relationship with synapses, cortex and serotonin (Cocchi et al. 2011b).

On the other hand, deformations of cell membranes under pathological conditions may be directly involved in the quantum aspect of consciousness according to the spin-mediated consciousness theory (Hu & Wu, 2004a, 2004b).

In other words, one could say that everything is relationship: that all reality is probably just interaction, the one in the other, as in our case, a conatus of mechanical nature that affects neuro chemical and neuroelectric realities. Ultimately it would be unavoidable the phenomenon of quantum consciousness in human and animal life, in its expression of classic consciousness (neurocorrelate) or confined to quantum consciousness (collective unconsciousness), for increasing levels, in response to the growing complexity of the animal world.

For the reasons discussed above, Linoleic Acid appears as a critical element of the symmetry breaking between major depression from one side and bipolar disorder, psychosis and normality, on the other side, practically, between the human and animal subject with MD and the rest of humans and animals. All the evidence emerged, to date, by the neuroscience studies, have failed to grasp exhaustively, at least in MD and BD, the most intimate mechanisms of the phenomena that affect the macro-areas of the brain. A decisive and perhaps even trivial ring failed in all the extraordinary complexity of the phenomenology of neuroscience. It has been underestimated, probably, the existence of an element, ie the mechanical force, intrinsic to the membrane.

In this regard, for example, different areas of the so-called " moral brain " have been identified (frontal and temporal lobes, cingulate cortex, amygdala, insula, hippocampus, basal ganglia, corpus callosum), however, remains obscure the matter of reflection and choice, as well as you can not tell whether the dynamics of the decision are attributable only to changes of recordable electrical activity of areas involved in these representations. **(Fumagalli and Priori 2012)**.

If we admit that the researches on the interactome are correct (there is no reason to suggest the contrary), if we assume that the difference in membrane mobility, to the point of no return, as in the case of MD, is the conditioning element which modify the functional expression of the interactome, then we must consider as decisive the mechanical force that the membrane exerts on the cytoskeleton, thus the deformation induced by different levels of mobility, to understand those changes that affect the neuro electrical and neuro chemical properties of neural structures. It is known, in fact, that modifications of mechanical pressure forces on the cytoskeleton induce electrical variations as well as those on ion channels, on the neurochemical side. Ultimately all the phenomenology leading to pathological expressions, at least in the case of mood disorders, would be induced by quantum phenomena. We can, now, rethink and redesign the meaning of the molecular hypothesis of consciousness (Figure 11).

In Figure 11 is described the molecular depression hypothesis made according to Cocchi and Tonello **(Cocchi and Tonello 2010a; Cocchi et al. 2010b)**, Rasenick **(Donati et al. 2008)**, Hameroff **(Hameroff and Penrose 1996)** experimental findings and the interactome relationships with quantum consciousness computation. The activation of PKC is preceded by a number of steps, originating from the binding of an extracellular ligand that activates a G-protein on the cytosolic side of the plasma membrane. The G-protein, using GTP as an energy source, then activates PKC via the phosphatidylinositol bisphosphate (PIP2) intermediate, which is shown as the DAG/IP3 complex. PKC activation activates adenylyl cyclase by promoting an increase of GTP. PKC and PKA, when activated, in turn, activate the AMPc resulting in modification of ion flux **(Gaginella and Galligan 1995; Albert 1998)**.

From the logic behavioral of the cell membrane, from the levels of symmetry breaking, would begin the long path of bio-molecular changes that, to adhere strictly to what is investigated, may affect consciousness, in mood disorders, through changes in mobility of the membrane, which the SOM so well has expressed and consistently interpreted. Cocchi and Tonello, Rasenick etc. show that differences were found in the membranes of patients with psychiatric disorders compared to supposed healthy subjects. Thus, it is plausible to assume that there are differences in the frequency of "conscious events" in patients with psychiatric disorders than in supposed healthy subjects (Figure 12).

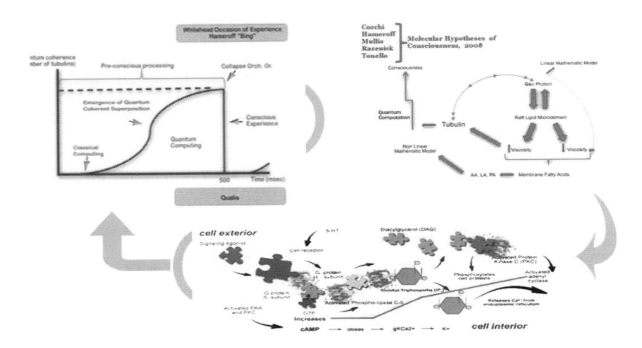

Figure 11. Schematic representation of the possible molecular and quantum consciousness connections.

In particular, Cocchi and Tonello have studied the Major and Bipolar Depressive Disorders. It seems plausible the hypothesis that there might be a lengthening of the of wave function collapse period (ie a reduction of frequency) in subjects with Major Depression (Figure 13), or that there might be a shortening of the same in subjects with bipolar disorder. It can be suggested a performance that alternates moments of high frequency with moments of low frequency (Figure 14). In Figure 15 is hypothesized the whole connection.

In short, the description of wave function collapses in Major Depression and Bipolar Disorder, in quantum consciousness, could be hypothesized as follow:

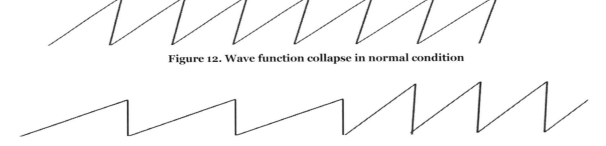

Figure 12. Wave function collapse in normal condition

Figure 13. Wave function collapse in depressive condition (lenthening alternating normality)

Figure 14. Wave function collapse in bipolar condition (shortening alternating lengthening)

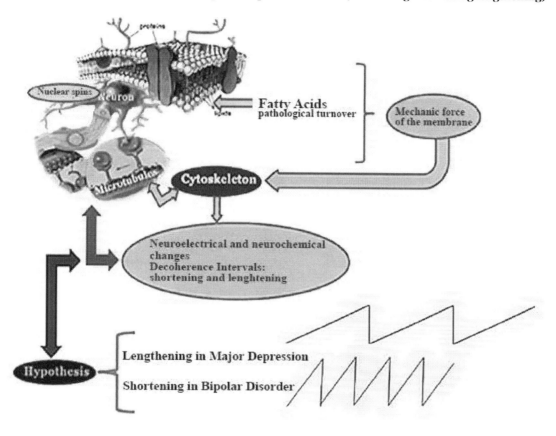

Figure 15. Hypothesis of connection among the mechanical force of the membrane, cytoskeleton, microtubules and nuclear spin, to explain the possible modification of the consciousness state.

Conclusion

The molecular sequence described and previously reported, which relates membrane mobility, protein Gsa and tubulin, would seem to be unavoidable in determining the axis of the micro molecular phenomena that allow the regulation of mood disorders. The research carried out on platelets and neurons, because of their similarity, led to the conclusive hypothesis that the above-mentioned axis could be responsible for the consciousness changes, and its modulation, and then, of the behavioral responses of humans and animals.

To conclude, we point out, therefore, the likely event that the neuro chemical and neuro electrical answer, from the functional point of view, in physiological and pathologic circumstances (namely MD and BD), may be due to modifications determined by the variations of the fatty acids dynamic of the phospholipid bilayer of neuronal membrane.

Finally, a brief mention of the "epistemological spirit" that has guided us: complexity *versus* ideological reductionism. We approached the consciousness processes, their dynamics in normal and pathological conditions, without staring on isolated elements and plans, as from a privileged and detached observatory, but thinking and acting operationally, on the merits of connections, relationships and networks.
All this, always in the belief that science is a story of an expressive world of emerging conditions and, for this reason, it can never be reduced to the trivial sum of its elementary ingredients and in which the symmetry breaking provide variety, creativity, vitality, as a sign of natural self-organization, of the evolving systems, thanks to spontaneous breakage of symmetry, to conditions of increasingly complexity and unpredictable.

As this process is relevant in the socio-economic context is self-evident: science embodied in economic and social structures, which, by accepting to transform from "foucoltians monitoring and control tools" (ideological reductionism) into open, sliding and emerging systems, (complexity or logic open), might actually, in the case of mental illness need to be aware of the unrealistic and confusing classifications of DSM and open connections between consciousness and quantum dynamics of the brain. In this sense, right from economic and social forces might result as Donald Mender, a psychiatrist at Yale, hopes: "The invention of quantum computers, which, although not yet practically in use, may supplant digital technology thanks to their purely computational power " (Licata and Sakaji 2010; Licata and Minati 2012; Anderson 1972).

References

Alberts B, Bray D, Lewis J, Raff M, Roberts K, Watson JD (1994) Molecular Biology of the Cell, 3rd edition, Garland Science, New York.
Allen JA, Halverson-Tamboli RA, Rasenick MM (2007) Lipid raft micro domains and neurotransmitter signaling. Nat. Rev. Neurosci 8: 128-140;
Anderson PW (1972) More is different, Science New Series, 177: 393-396.
Apodaca G (2002) Modulation of membrane traffic by mechanical stimuli. AJP-Renal Physiol 282: F179 F190.
Austin MP, Ross M, Murray C, O'Carroll RE, Ebmeier KP, Goodwin GM (1992) Cognitive function in major depression. Journal of Affective Disorders, 25: 21–29.
Barzanti V, Battino M, Baracca A, Cavazzoni M, Cocchi M, Noble R, Maranesi M, Turchetto E, Lenaz G (1994) The effect of dietary lipid changes on the fatty acid composition and function of liver, heart and brain mitochondria in the rat at different ages. The British journal of nutrition 71 (2).
Basso MR, Bornstein RA (1999) Relative memory deficits in recurrent versus first-episode major depression on a word-list learning task. Neuropsychology 13: 557-63.
Bearden CE, Glahn DC, Monkul ES, Soares JC, Barratt J, Najt P, Kaur S, Sanches M, Villareal V, Bowden C (2006) Sources of declarative memory impairment in bipolar disorder: mnemonic processes and clinical features. Journal of Psychiatric Research 40: 47–58.
Benedetti S, Bucciarelli S, Canestrari F, Catalani S, Cocchi M, Colomba MS, Gregorini A, Mandolini S, Marconi V, Mastrogiacomo AR, Rasenick M, Silvestri R, Tagliamonte MC, Tonello L, Venanzini R (2012) Molecular Changes in Mood Disorders Results of the Marche Region Special Project, NeuroQuantology. 10: S1-28.
Bok E, Hryniewicz-Jankowska A, Sikorski AF (2009) The interactions of actin cell and membrane skeleton proteins with lipids. Postepy Biochem 55: 207-22.
Bourre JM, Piciolti M, Dumont O, Pascal G and Durand G. (1990) Dietary Linoleic Acid and Polyunsaturated Fatty Acids in Rat Brain and Other Organs. Minimal Requirements of Linoleic Acid. LIPIDS 25 (8).

Breslow R, Kocsis J, Belkin B (1980) Memory deficits in depression: Evidence utilizing the Wechsler Memory Scale. Perceptual and Motor Skills 51: 541–542.

Bruza P. Idealistic quantum psychopathology, NeuroQuantology. 2010, 8: Page 64-65;

Burt DB, Zembar MJ, Niederehe G (1995) Depression and memory impairment: a meta-analysis of the association, its pattern, and specificity. Psychological Bulletin, 117: 285–305.

Chen W, Jump DB, Essekman WJ, Busik JV (2007) Inibition of cytokine signalling in human retinal endothelial cells through modification of caveolae/lipid rafts by decosaexaenoic acid. Invest. Ophthalmol. Vis Sci 48: 18-26.

Chifflet S and Hernandez JA (2011) The PlasmaMembrane Potential and the Organization of the Actin Cytoskeleton of Epithelial Cells, International Journal of Cell Biology Volume 2012 Article ID 121424, 13 pages, doi:10.1155/2012/121424.

Cocchi M et al. Major Depression from Symmetry Breaking. NeuroQuantology, in press;

Cocchi M and Tonello L (2006a) Biological, Biochemical and Mathematical considerations about the use of an Artificial Neural Network (ANN) for the study of the connection between Platelet Fatty Acids and Major Depression. J Biol Res LXXXI, 82- 87.

Cocchi M, Tonello L, Cappello G (2006b) Biochemical Markers in Major Depression as interface between Neuronal Network and Artificial Neural Network (ANN). J Biol Res 81: 77-81.

Cocchi M and Tonello L (2007a) Platelets, Fatty Acids, Depression and Cardiovascular Ischemic Pathology. Progress in Nutrition 9: 94-104.

Cocchi M and Tonello L (2007b) Depressione Maggiore e Patologia Cardiovascolare Ischemica. Un Network fra piastrine, cuore e cervello. Reti Neurali Artificiali, potenzialità predittiva di alcuni acidi grassi della membrana piastrinica. Bologna, CLUEB Editore.

Cocchi M, Gabrielli F, Tonello L, Pregnolato M (2010b) The Interactome Hypothesis of Depression. NeuroQuantology 4: 603-613.

Cocchi M, Gabrielli F, Tonello L, Pregnolato M. (2011a). Consciousness and Hallucinations: Molecular Considerations and Theoretical Questions. NeuroQuantology 9: 182-189.

Cocchi M, Marzona L, Pignatti C, Olivo OM (1979) Effect of organ phospholipids on the growth of embryonal tissues cultured in vitro. Biochem Exp Biol 15: 13-6.

Cocchi M, Sardi L, Tonello L, Martelli G (2009d) Do mood disorders play a role on pig welfare? Ital J Anim Sci 8: 691-704.

Cocchi M, Tonello L (2010a) Bio molecular considerations in Major Depression and Ischemic Cardiovascular Disease. Central Nervous System Agents in Medicinal Chemistry 10: 97-107. Cocchi M and Tonello L (2012) How mathematics can inform the diagnosis of Mood Disorders. NeuroQuantology 10: S1-28.

Cocchi M and Tonello L (2010b) Running the hypothesis of a bio molecular approach to psychiatric disorder characterization and fatty acids therapeutical choices, Annals of General Psychiatry 9 (supplement 1): S26.

Cocchi M, Tonello L, De Lucia A, Amato P (2009a) "Platelet and Brain Fatty Acids: a model for the classifcation of the animals? Part 1". International Journal of Anthropology; 24: 69-76.

Cocchi M, Tonello L, De Lucia A, Amato P. (2009b) Platelet and BrainFatty Acids: a model for the classification of the animals? Part 2. Platelet and Brain Fatty acid transfer: Hypothesis on Arachidonic Acid and its relationship to Major Depression, International Journal of Anthropology 24: 69-76.

Cocchi M, Tonello L, Gabrielli F (2012). The animal side of "mood disorders". Saarbrücken, Germany: LAP Lambert Academic Publishing.

Cocchi M, Tonello L, Gabrielli F, Pregnolato M, & Pessa E (2011b) Quantum Human & Animal Consciousness: A Concept Embracing Philosophy, Quantitative Molecular Biology & Mathematics. Journal of Consciousness Exploration & Research 2: 547-574.

Cocchi M, Tonello L, Tsaluchidu S, Puri BK (2008) The use of artificial neural networks to study fatty acids in neuropsychiatric disorders. BMC Psychiatry 8 (Suppl 1): S3.

Cocchi M, Tonello L, Rasenick Mark M (2010c) Human depression: a new approach in quantitative psychiatry. Annals of General Psychiatry 9:25.

Colby CA, Gotlib IH (1988) Memory deficits in depression Cognitive Therapy and Research, 12: 611–627.

Deckersbach T., Savage CR, Reilly-Harrington N, Clark L, Sachs G, Rauch SL (2004) Episodic memory impairment in bipolar disorder and obsessive–compulsive disorder: the role of memory strategies. Bipolar Disorders 6: 233–244.

Donati RJ, Dwivedi Y, Roberts RC, Conley RR, Pandey GN, Rasenick MM (2008) Postmortem brain tissue of depressed suicides reveals increased Gs alpha localization in lipid raft domains where it is less likely to activate adenylyl cyclase. J Neurosci. 28: 3042-3050.

Fossati P, Harvey PO, Le Bastard G, Ergis AM, Jouvent R., JF (2004) Allilaire Verbal memory performance of patients with a first depressive episode and patients with unipolar and bipolar recurrent depression. Journal of Psychiatric Research 38: 137–144.

Fumagalli M, Priori A (2012) Functional and clinical neuroanatomy of morality. Brain, A Journal of Neurology 135: 2006-2011.

Gaginella TS and Galligan JJ (1995) Serotonin and gastrointestinal function, CRC Press, USA.

Garcia-Sevilla JA, Padro D, Giralt MT, Guimon J, Areso P (1990) Alpha2-adrenoceptor-mediated inhibition of platelet adenylate cyclase and induction of aggregation in major depression. Effect of long term cyclic antidepressant drug treatment. Arch Gen Psychiatry. 47: 125–132.

Golinkoff M, Sweeney JA (1989) Cognitive impairments in depression. Journal of Affective Disorders 17: 105–112;

Green P, Gispan-Herman I, Yadid G (2005) Increased arachidonic acid concentration in the brain of Flinders Sensitive Line rats, an animal model of depression. J Lipid Res 46: 1093-6.

Hamazaki K, Hamazaki T, Inadera H (2012) Fatty acid composition in the postmortem amygdala of patients with schizophrenia, bipolar disorder, and major depressive disorder. Journal of Psychiatric Research 46: 1024-1028.

Hameroff SR, Penrose R (1996) Orchestrated reduction of quantum coherence in brain microtubules: a model for consciousness. In Toward a Science of Consciousness - The First Tucson Discussions and Debates Edited by: Hameroff SR, Kaszniak A, Scott AC. Cambridge, MA, USA: MIT Press. pp 507-540.

Hines LM and Tabakoff B (2005) Platelet adenylyl cyclase activity: a biological marker for major depression and recent drug use. Biol Psychiatry. 58: 955–962.

Hoffman PL, Glanz J, Tabakoff B (2002) Platelet adenylyl cyclase activity as a state or trait marker in alcohol dependence: results of theWHO/ISBRA Study on State and TraitMarkers of Alcohol Use andDependence. Alcohol Clin Exp Res 26: 1078–1087.

Hoover RL, Fujiwara K, Klausner RD, Bhalla DK, Tucker R, Karnovsky MJ (1981) Effects of Free Fatty Acids on the Organization of Cytoskeletal Elements in Lymphocytes, MOLECULAR AND CELLULAR BIOLOGY 10: 939-948.

Horan WP Poggee DL, Borgaro SR, Stokes JM, Harvey PD Learning and Memory in Adolescent Psychiatric Inpatients with Major Depression: A Normative Study of the California Verbal Learning Test (1997) Archives of Clinical Neuropsychology. 12: 575 584.

Hu H & Wu M (2004a) Spin-mediated Consciousness Theory: Possible Roles of Neural Membrane Nuclear Spin Ensembles and Paramagnetic Oxygen. Medical Hypotheses, 63: 633-646.

Hu H & Wu M (2004b) Action Potential Modulation of Neural Spin Networks Suggests Possible Role of Spin in Memory and Consciousness, NeuroQuantology, 2: 309-317.

Janmey P. (1995) Cell Membranes and the Cytoskeleton, Chapter 17, Handbook of Biological Physics, Volume 1, edited by R. Lipowsky and E. Sackmann, Elsevier Science B.V.

Janmey PA and McCulloch CA (2007) Cell Mechanics: Integrating Cell Responses to Mechanical Stimuli, Annu. Rev. Biomed. Eng 9:1–34.

Kohonen T (2001) Self-Organizing Maps. 3. Berlin: Springer.

Li R and Bowerman B (2010) Symmetry Breaking in Biology. Cold Spring Harb Perspect Biol 2: a003475.

Licata I, Minati G (2012) Meta-structural properties in collective behaviours. International Journal of General Systems 41: 289-311

Licata I, Sakaji A eds. (2010) Crossing in Complexity: Interdisciplinary Application of Physics inBiological and Social Systems, Nova Science Publishers. ISBN 9781616680374.

Maldjian A, Farkas K, Noble RC, Cocchi M, Speake BK (1995) The transfer of docosahexaenoic acid from the yolk to the tissues of the chick embryo. Biochimica et biophysica acta 1258(2).

Marei WF, Wathes DC and Fouladi-Nashta AA (2010) Impact of linoleic acid on bovine oocyte maturation and embryo development. Reproduction 139: 979–988;

McIntosh TJ, Simon SA (2006) Roles of bilayer material properties in function and distribution of membrane proteins. Annu. Rev. Biophys. Biomol. Struct. 36: 177–98.

Menninger JA and Tabakoff B (1997) Forskolin-stimulated platelet adenylyl cyclase activity is lower in persons with major depression. Biol Psychiatry 42: 30–38.

Mooney JJ, Samson JA, McHale NL, Colodzin R, Alpert J, KoutsosM, Schildkraut JJ (1998) Signal transduction by platelet adenylate cyclase: alterations in depressed patients may reflect impairment in the coordinated integration of cellular signals (coincidence detection). Biol Psychiatry 43: 574–583.

Mooney JJ, Schatzberg AF, Cole JO, Kizuka PP, Salomon M, Lerbinger J, Pappalardo KM, Gerson B, Schildkraut JJ (1988) Rapid antidepressant response to alprazolam in depressed patients with high catecholamine output and heterologous desensitization of platelet adenylate cyclase. Biol Psychiatry 23: 543–559.

Namikoshi M, Suzuki S, Meguro S, Kobayashi H, Mine Y and Kasuga I (2002) Inhibitors of Microtubule Assembly Produced by the Marine Fungus Strain TUF 98F139 Collected in Palau. Journal of Tokyo University of Fisheries 88: 1-6.

Noble RC, Cocchi M (1990) Lipid metabolism and the neonatal chicken. Progress in lipid research 29 (2).

Pandey GN, Pandey SC, Janicak PG, Marks RC, Davis JM (1990) Platelet serotonin-2 receptor binding sites in depression and suicide. Biol Psychiatry 28: 215–222.

Popova JS, Greene AK, Wang J, Rasenick MM (2002) Phosphatidylinositol 4, 5-bisphosphate modifies tubulin participation in phospholipase Cβ1 signaling. J. Neurosci 22: 1668-1678.

Rasenick MM, Donati RJ, Popova JS, Yu JZ (2004) Tubulin as a regulator of G-protein signaling. Methods Enzymol 390: 389-403.

Sheetz MP, Sable JE, and Döbereiner HG (2006) Continuous Membrane-Cytoskeleton Adhesion Requires Continuous Accommodation to Lipid and Cytoskeleton Dynamics. Annu. Rev. Biophys. Biomol. Struct. 35: 417–34.

Sottocornola E, Berra B (2008) Modulation of membrane lipid raft by omega-3 fatty acids and possible functional implication in receptor tyrosinekinase activation. Progress in Nutrition 10: 210-213.

Stromgren S (1977) The influence of depression on memory. Acta Psychiatrica Scandinavica 56: 109–128.

Stulnig TM, Huber J, Leitinger N (2001) Polyunsaturated eicosapentaenoic acid displaces proteins from membrane rafts by altering raft lipid composition. J Biol Chem 276: 37335-40. Han X, Smith NL, Sil D, Holowka DA, McLafferty FW, Baird BA (2009) IgE receptor-mediated alteration of membrane-cytoskeleton interactions revealed by mass spectrometric analysis of detergent-resistant membranes. Biochemistry 48: 6540-50.

Sun M, Northup N, Marga F, Huber T, Byfiel dFJ, Levitan I and Forgacs G (2007) The effect of cellular cholesterol on membrane cytoskeleton adhesion, J Cell Sci 120: 2223-31.

Svennerholm L (1968) Distribution and fatty acid composition of phosphoglycerides in normal human brain. Journal of Lipid Research Volume 9.

Tonello L and Cocchi M (2010) The Cell Membrane: Is It A bridge from psychiatry to quantum consciousness? NeuroQuantology 8: 54-60.

Yap B, Kamm RD (2005) Cytoskeletal remodeling and cellular activation during deformation of neutrophils into narrow channels. J Appl Physiol 99: 2323–2330.

Foundation of Reality: Total Simultaneity

Wilhelmus de Wilde[*]

ABSTRACT

In this article, I put forward the hypothesis of Total Simultaneity ("TS"), a "fifth" dimension behind the Wall of Planck reality at which we pass the limits of causality at the quantum scale and "Now" as we perceive no longer exists. TS can be reached by every point of our 4 dimensional universe and singularities only exist in our consciousness. The other limit of causality is the local speed of light c at which time stands still so there is no more before and after. All information of all parallel universes and multiversity constitution in TS is simultaneously present and available, our consciousness is able to align points out of the TS and so create the observable analogue universe that we are aware of. The totality of information from other universes (also partly observable by other consciousness) is influencing our linear causal deterministic universe, the origin of gravity, dark matter, and the dark energy may emerge from here. The Big Bang is an imaginary non-existing point in the TS area. Inflation is avoided by projecting inflation time into the area after the Wall of Planck, uniformity in the structure of space-time is also guaranteed. Our mind with its 100 billion neurons is able to cope with infinities because it has parallels with the qualities of TS.

Key Words: reality, simultaneity, fifth dimension, Planck, wall.

The Causal Limits

One night when I came home late at night, I left my car and contemplated the night sky with its beautiful view on the stars and our milky way; (living in the country means also that there is no trash light disturbing the view). I recognised the stellar formations that mankind already recognise since the early Egyptians, the imaginary lines that form imaginary drawings in the sky. But can't we also draw these imaginary lines in different ways so that other forms and constellations are created? The creative freedom of our minds has no limits.

While looking at this marvellous lightshow, I got the idea that each line between these points of light could be interpreted as a possible reality sequence of another universe, and that all these realities are already present, you only have to draw a line between points of light. Drawing such a line is the typical way of expressing history in our causal four-dimensional world, where the arrow of time has a direction from past to "now" to the future.

Describing the multiversity and parallel worlds can only be incomplete, because we have no means to describe other dimensions, especially one where all the pasts, the now's and the futures

---

[*] Correspondence: Wilhelmus de Wilde. E-Mail: wilhelmus.d@orange.fr Note: This article was first published in Scientific GOD Journal, 2(4): pp. 334-342 in 2011 and is based on a submission to FQXi Essay Contest " Is Reality Digital or Analog?" in 2011.

of all possible universes are united, you can call it a "fifth" dimension at the edges our own but this is also a sequential expression a down to earth explanation.

One of these edges is what scientists describe as the "Wall of Planck" (10^{-33}cm / 10^{-43}sec), behind this Wall we encounter the creativity of humans, like M theory: 10 enrolled dimensions and/or branes and on the other hand we encounter singularities with no dimensions at all. These theories can (until now) not be proved with tests, science becomes philosophy with presumptions, hypotheses, assumptions and countless beautiful formula's that are very difficult to understand for non specialists , also metaphysical and transcendental ideas are soliciting on an equal base for a place in this area.

So, once we reached this Wall of Planck, behind it we would experience the non causal dimensions of the origin of our own space-time and many other universes, which also means that after this Wall there is **no separate past, no separate now and no separate future. It is the All in One, the <u>Total Simultaneity</u>** (from now on to be called: TS, not to be confused with the Absolute Simultaneity of Albert Einstein that is a simultaneity occurring in a causal relativistic universe) where all possible pasts, now's, futures and places of all thinkable and non-thinkable universes are simultaneously "present", comparable with our memory where all the events of the past have an **equal place,** only active thinking replaces this events in a linear causal sequence. The TS is what we will refer to as a fifth omnipresent dimension.

We can in fact reach it at any point in our space/time continuum by approaching the Wall of Planck, until now we cannot yet reach behind this wall, so the "constitution" of this dimension is still unknown, every part behind the Wall of Planck is of the "same quintessence" (in reference to Aristotle where he cited that the world was not only made of earth, water, air and fire but also of a fifth element that made it possible to function) with the same properties, so it is as we can see our universe emerge from this fifth dimension, every space/time quantum having a different quality and quantity of information retrieved from TS. The diameter of these quanta (spheres) is the Planck-length, on a bigger scale the universe that descends from this dimension is appearing as a continuum obeying to thermodynamic laws, that differs totally from the quantum laws, being the reason why **<u>General Theory of Relativity and Quantum physics will never merge to a unity.</u>**

When we apply "Ockham's Razor" this approximation is a simpler approach of the origin and existence of our universe, we have no longer to deal with infinities that will lead to paradoxes in our linear world. There is no need for multiple dimensions/branes and/or points without any dimension (every one of them also "existing" after the Wall of Planck), later on we will come back to the origin of the existing fine-tuned constants.

Our Consciousness

Imagining a line between certain possible points in the TS makes our consciousness understand history and the qualities of our universe. All sort of possible imaginary lines, between the infinity of information points available, (information is of another quality as data) can be drawn for all other imaginary and non-imaginary universes. It is our consciousness that "connects" these points and creates the lines that we perceive, like a piece of music that becomes "reality" in our mind by connecting points of our memory and so forming a sequence.

We must also apply this view-point to the cores of black holes that should contain a singularity, but singularities are points inside the area of TS, we only have to realise that each point in our universe touches this Wall of Planck and so TS, replace singularities by the quantum with the diameter of 1.616252×10^{-33} cm, at the edge of TS. The information of this core-quantum equals the information of the quanta of the beginning of our universe.

Creation of our universe ex-nihilo is no longer needed, creation of something out of nothing, singularities, FIAT LUX as proposed by our religions, our causal material universe does not allow infinities, because there are borders. Creation is in fact all around us, every moment. Transcendent theories are explaining the beginning by assuming an external "GOD" as a Creator and all the other infinities that the human spirit/mind is able to think of. Aren't we are obliged to divide our infinities in TWO in order to understand a Beginning and an End, a Yes and a NO, we have to digitalise them. Once an infinity is divided in two it is no longer an infinity, but is it two infinities? Paradoxes all over, these occur from the difference in structure of our "multi-reality TS" minds and the analogue causal deterministic reality world that we live in, where the edge on the quantum side is the Wall of Planck.

We are at this moment in the proficiency to "create" for two of our senses, sight and hearing, a digital world, constituted of 0's and 1's, or should I say Zero's and Ones, or is it better even for the understanding to name them Yes or No, to indicate the question(s) before. These CAUSAL SEQUENCES of two understandings that constitute so called soft-ware that with the aid of a complex machine called computer can let emerge for our eyes and ears a world of video and sound are the basics of our digital reality. In our consciousness we replace ourselves in this virtual digital reality, for example in a role playing game, the bearers of these data (in the role playing game there are different ways to experience an adventure called life) has some parallels with the TS, different lines of reality can be followed, lining up different quanta of information of the time/space continuum. However the exactness of the sequence of these yes and no's, in our digital causal world is of the essence, if one zero is replaced by a one then the whole perception falls apart the virtual reality becomes blurred, this is a typical property of our causal four dimensional world.

Bits, Qubits and Reality

In the world of the future quantum computer, qubits represent the equivalent of the Yes and NO. However qubits don't offer only two possible choices of yes or no, each qubit in fact represents an infinite superposition of possibilities, as indicated in the "BLOCH" sphere (the superposition of the two states is described by a linear combination with the form $a \times 0 + b \times 1$, where all the values of a and b are complex numbers) all these possible quantum states of a qubit we have to bring back to our "digital" status of 1 or 0, because we have to experience the results as causal sequences.

The qubit's superimposed state is comparable with the infinite possibilities of drawing reality-lines in the TS. In fact, once we will have constructed a quantum computer, even when this "machine" is not "working" (under tension), it will hold already all the answers for all our possible questions also those that not yet posed. So we have to try to search for a more adequate form of "handling" in order to optimise our results, maybe the future solution will not be the way used in the binary computer (for instance: for 1000 qubits we have 2^{1000} possible configurations: 10^{300} which is more as all the atoms in our universe, so we can stock 10^{300} solutions for a problem, in other words we can treat at the same time 10^{300} potential solutions.) Four stages of "reality" are appearing:

1. On the smallest scale the "**quantum" multi-reality TS**" which is a superposition of states, out of this multi-reality this quintessence contains the information of the quanta spheres with diameter the Planck length that are by our consciousness unified into linear sequences constituting our analogue "reality". When we are constructing a quantum computer it should be possible to let emerge a new analogue reality, like forming a new reality-line in the TS that should also be accessible for our consciousness.

2. Our **analogue reality** "the every day" reality which is the result of the for our consciousness sensible line in the TS (see 1). In this reality, each human is aware of his own sequences of points in the Total Simultaneity, comparable with the phenomenon that our consciousness perceives a film on television while in fact it is a sequence of digital bits.

3. The **digital virtual reality**, which is a sequence of bits that is created by ourselves (until now for two senses), because of the causal property of these sequences (it is created in our four-dimensional causal deterministic world) this virtual reality will be totally dependent of the reality as experienced in our consciousness and so is a sub-reality.

4. The "**social" analogue reality** that is formed by the human "qubit", forming society's that seem to have an existence of their own, but in fact are receiving the input from the single bits of information, this social reality has also parallels with the quantum reality because of the fact that every qubit has multiple information available like a superposition of states. We are now with about 6,9 billion people on earth. Every human is a bearer of information, like a qubit, together we are forming a complex system reacting as a whole, making progress in science as a whole, making war as a whole, especially in this information era every "qubit" of mankind reacts(output) directly on events (input), thus transforming his observable analogue reality and creating a over-consciousness (not sub).

In the future digital virtual linear reality's will be created for our five senses in which mankind can experience virtual multiverses, imagined by himself, attaining another level of consciousness, a sub reality like it is formed by the sequences in the T S. Shall we create a new kind of digital "consciousness" a sub-consciousness linked to our present consciousness that is aware of this virtual world as if it was its real analogue world? Will it be possible to create for this sub-reality a new form of consciousness that like our consciousness will start searching for his own grail of understanding that on its turn, created him. Will this sub-consciousness once also reach to the echelon of a non-linear universe?

Continuous Creation of Reality

Why is it that our consciousness is sensible for information available in the TS and is able to "handle" it? Our mind is made of neurones, each human has 100 billion neurones (!) where electrons are moving from one neuron to another, if these electrons were in superposition (wave or particle) they would encounter at every split of neurons a decision point (like the Young Double Slit experiment) spiralling from one information point in TS to another and thus creating for our consciousness a analogue causal linear universe.

All together on one side our five senses are giving the input of the signals we receive (only signals from the past) , thus these observations changing into information, this information is brought back to electrical currents in our super (quantum) computer of 100 billion neurones, creating all kind of possible observable realities, (see page 3 where 1000 qubits are available for 10^{300} solutions!) one of which is taken as the analogue sensible reality that someone will be aware of.

This newly created virtual world also has to obey to specific fine-tuned constants (exact sequences), if one of them differs from the constants as we perceive them, that specific sub-reality will longer exist for us. You can wonder if we will be able to create the kind of exactness that is needed in arranging these specific sequences that are in conformity with the exactness of all the fine tuned constants that we perceive around us (Cosmological, alpha, value and charge of the electron....).

When approaching this problem it is like solving the problems of so called Quasi Polynomial Time and NP complete decision problems, it would be adequate to be able to go back in time in order to realise "more" time. This was the origin of the so called time-loops (David Deutsch, Oxford University, 1991), these time loops make space-time interchangeable with the parallel universes in order to make them accessible for calculation, the grandfather-son paradox that will not occur because we arrive at another line of "reality" **in TS, we are entering in a SPIRAL timeline.**

Our quantum computer as a tool is able to, before showing solutions, in a superposition of states, like the TS, create possibilities like using parallel time and parallel consciousnesses'. **By reaching out at the qualities of the TS, we may be able to touch in this way the dimension/area after the Wall of Planck.**

In our analogue world we encounter different forces and one of the forces to enhance into the Theory of Everything is the gravitational force. Scientists are trying hard to understand it, one of the theories that come near to the TS notion is the theory posted by Eric Verlinde of the University of Amsterdam ,(see also FQXi: The myth of gravity, 01-10-2011) he draws an analogy between thermodynamics and gravity. The effects attributed to gravity can thus be described as results from forces that are on the edge of our universe, (in the TS Theory this edge is the Wall of Planck). On our scale (analogue) the perceptible world is constituted of sequences of billions of quanta forming a "unity", this unity has whole obeys to different "laws" as the quanta scale that constitutes it. Verlinde describes our universe as a hologram and at the border of this frontier. The TS viewpoint states that this 4-dimensional hologram is to be refreshed every 5.39121×10^{-43} sec (Planck time). The sequence of refreshment so creates for our consciousness's again the linear causal time. So all the effects like Gravity but also themes like the Dark Matter and the Dark Force that we have to deal with, can change in disposition and laws on greater scales and in smaller scales.

"In the beginning was the world, and the world was LOGOS"

What we perceive as the baryonic particles that represents only ca. 5% of our "visual" baryonic material, is the sequential causal effect of the "fifth" dimension. **It is too narrow minded to expect that if we are aware of only 5% (or even less) of the universe, the other 95% has to obey the same Laws and be consisted of the same kind of particles**. It is more understandable that our neighbours are influencing us the results of their actions can be observed.

We can understand in our minds why entangled particles when observed don't respect the velocity of light (velocity is a causal sequence in our four-dimensional world). They are in a superposition in TS, once we observed one of them in our causal world it seems as if simultaneously his brother takes the same state, comparable with the Young two slit experiment, because two and more realities exist already simultaneously in TS, it is as if the total of possibilities can pass by the fence with one or two slits, all of them ready to become for us "reality", at a distance of 1.616252×10^{-33} cm before the slit(s). Then one of the imaginary lines in TS is becoming "reality". The observer is able to realise/construct one or two slices in his causal world before the last 5.39121×10^{-43} sec before the moment that the photon/wave function is in front of the barrier.

The place/moment 10^{-33} cm/10^{-43} sec before the slit(s) has the same "quality" as the imaginary point Zero. Once the consciousness of the observer realises an observation (the observer in our case is constituted of baryonic particles), his observation creates out of the observed entanglement, (the superposition of the wave function) an "observable" particle that has to be a baryonic particle; otherwise, he would not be able to observe it. When "we" are observing the "face of God", (George Smoot 1992), we are observing the moment that first light was emitted 380.000 years after the imaginary point Zero, at that very moment the observer is **materialising himself** and his whole perceptible universe, he is not only looking in the MIRROR but at the same time creating this mirror.

For these observed entangled photons to whom time did not pass,(because they travelled at the speed of light in our 4-dimensional world) , it was as if they encountered our eyes at the same time they left there entangled particles (brothers) in the primary "soup", these partners at that very moment acquired the same for us observable properties, it is like the old myth of the orobouros, the snake that eats his own tail. **We are the origin of our own universe,** our perceptible linear universe, no more problems with its specialities, its uniqueness , no creationism needed in the way that there was an external GOD who created this world especially for the human kind (this is one of the results of the "social analogue reality"), it is the human consciousness (LOGOS) that creates his own universe as a unique universe for himself but this is only one of **an endless row created by other consciousnesses , every one of them perceptible by his own "observers" or creators.**

The Great "Beginnings & End" Conclusion

The assumption until now was that going back in time for the search for the beginning of our universe is a linear one, we thought that we ought to begin at Zero, from the so called Big Bang to the NOW. The interpretation of the cosmic background radiation with its temperature of 2,725°K was one of the reasons that obliged us to draw that line back to ZERO . This point cannot be reached however because its is a imaginary point in the TS , our journey to this point can not be achieved , so **there is no longer a singularity to be explained , neither there was a Big BANG.**

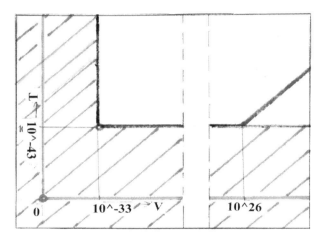

In order to visualise this, we imagine the timeline as a vertical one and the volume of space as a horizontal line, we meet on the cross point of the timeline and the space-line our point ZERO, that was called the BIG BANG. On the time line we move up to the point 1.616252×10^{-33}sec and on the space line we move to the right to the point 5.39121×10^{-43}cm, at the cross point of the lines , every point at the left side of the new time-line and under the new space-line is in TS area. So we see that the former zero point does not exist anymore as a causal sequential point (see illustration).

On the horizontal line of 10^{-43}sec we will then find a point with a, for instance, volume of 10^{26}cm³ that is in conformity with the volume of our universe after the "inflation" (as perceived by now), thus accounting for the volume that our causal sequential universe contained Any volume could emerge to reality. This "emerging" of the universe at the time of 10^{-43}sec at a volume of 10^{26}cm³ can be understood as the moment that an observer received this partial information out of TS "regarding" the volume of 10^{26} cm³ at that specific moment being 10^{-43} sec. At that point, the emerging of our universe can be seen as a linear one on the edge of the Planck time line.

What we do is replace the inflation time as presumed to be from 10^{-35}sec till 10^{-32}sec on the line of the 10^{-43}sec, so this inflationary time is extracted from the TS "time". **This so called Inflationary time is no longer existing in our sequential causal universe**, uniformity in the structure of the universe has become a simple logic, the mirror has become a linear causal reality, emerged from TS where all possible information has a potential to become observable for future observers, whatever the structure of the particles their universe is constituted of.

Our perceptible universe is positioned in a **time-loop** that begins its linear sequential existence 5.39121×10^{-43} sec after a imaginary point Zero, with all conditions present that made it possible for our human consciousness to exist, as one of the multiple possibilities/information lines existing in the TS. This time-loop becomes a spiral one when different observations are made. We than enter new parallel universes (the old one is still existing), the only difference with theories of today is that all information regarding these parallel worlds is always permanent existing and we merely jump over to another "information-point" in TS.

Accepting a beginning that emerges with our observation from the TS, which can be seen as the quintessence where all possible information is stocked, only not in a causal linear way, becomes more comprehensible. Of all the information available, we only align in our consciousness a tiny special part, the splitting up of the universe every time that a decision is made and the mass/energy needed for the creation of these universes is no longer needed, we only touch another information point in the TS (like the available information stocked on a hard-disk) thus opening a new line of possibilities.

Because of the parallels existing in the structure of our mind and in the structure of TS, the lines as they are observed by our consciousness in TS, in a way influence each other, our consciousness becomes aware. The special line formed by our consciousness is forming our lives the world around us becomes "reality". We think that we perceive and understand the analogue reality around us, the exactness of the parallel approach of the two "lines" in separate dimensions (the four dimensions of our linear deterministic world and the fifth dimension of TS) and the way they that influence each other will perhaps never be totally understood.

Our minds only will have symbols and mathematical equations to express our infinities. It is true that the human consciousness enables us to think of and deal with all these external infinities, we can imagine fractal universes with endless borders, while in our analogue universe these infinities will only lead to paradoxes. Mankind "feels" however this infinite TS presence, not as a pure physical phenomenon but as a "spiritual" experience. Since the beginning Myths and Legends of other worlds accompany us, religion is one of the pillars to understand our universe it is like the Theory of Everything that scientists are looking for. The human mind however "believes", and these beliefs emerge as the fourth reality the social reality.

Our "touchable" analogue causal reality is only one of the many realities which is originally formed in our consciousness. We are in the process now of creating a sub-reality in the form of a digital one dependent of and emerging from a linear causal interpretation, this can never become or even have the same quality of this first reality. However it should be possible to create a consciousness, not the sub-consciousness above mentioned, but a parallel one that can be linked to our own, also aware of all different fine-tuned Constants. Parallel worlds could become perceivable and understandable this could open even communication with the newly created consciousness as interpreter.

Our own special multi-reality TS 100 billion neurones mind, adequate to be the origin of his own universe, will in the future evaluate further and further, the information stocked up in TS will never be totally admissible. The key of our minds will be able to open this until now dark side of our universe. It will be the password to open other parallel universes, newly structured quantum computers can become A reality, thus opening all the chances available on the other side, but most of all we will be able to understand more deeply our existence/reality whether it is digital or analogue, in the meantime we will admire the nightly sky.

References

David Deutsch, Quantum Mechanics near closed timelike lines, Oxford University 1991

Richard P. Feynman, The Character of Physical Law (with introduction of Paul Davies), Penguin Books 1992

George Smoot, 23 April 1992, commenting the images of the COBE satellite: "If you are religious, it is like looking at GOD.

Alan Guth, Het Uitdijend Heelal (original : The Inflationary Universe ,1997) , Uitgeverij CONTACT, Amsterdam/Antwerpen , 1998.

Brian Greene, L'Univers Elégant 5the Elegant Universe, Editions Robert Laffont, S.A. Paris 2000.

Trinh Xuan Thuan, Les Voies de la Lumière ; FAYARD, 2007.

Lee Smolin, Rien ne va plus en Physique (The trouble with Physics - The Rise of String Theory, the Fall of a Science and What Comes Next, 2006) DUNOD, Paris 2007.

Leonard Susskind, Le Paysage Cosmique (The Cosmic Landscape), Editions Robert Lafont, S.A. Paris, 2007.

Etienne Klein, Discours sur l'Oribine de l'Univers. Flammarion 2010.

Igor and Grchka Bogdanow, Le Visage de DIEU, Editions Grasset & Fasquelle 2010.

Eric Verlinde, The Myth of Gravity : FQXi Article , January 10, 2011

A Metaphysical Concept of Consciousness

Wilhelmus de Wilde[*]

ABSTRACT

Our five senses ask for the reference of reference, which cannot be found in our causal material reality. We all have a subjective reality that is presented as a simultaneity sphere around our consciousness. Consciousness not being a material entity with length or volume can be treated as a singularity. The cutting circles of the different spheres around the percipients form a foam (origin of decoherence) of objective mutual reality (history). Our Reductionist Causal Deterministic Universe (RCDU) is limited by the Planck Wall. Before that causality doesn't exist, we call this "dimension" Total Simultaneity (TS). The causal time-line (β-time) consists of entangled Alpha–probabilities (α-P's). Each "observer" is the origin of his own causal time lines. Free will makes the choice out of infinity of α-P's in TS. Subjective individual time travel can happen without being conscious of the process (until now). We explain the double-slit experiment, the spooky action at a distance, the Many World Interpretation, Entropy, Time travel and free will. We indicate that the reductionist way of interpretation is only valid for a limited part of our reality and may not lead to the "truth" finding of the whole. The emergent way of approach may be more effective and create an understandable "being" of WHAT IS. (Parmenides, Greek Eleatic School of Philosophy, 515 BC).

Key Words: reality, simultaneity, fifth dimension, Planck Wall.

The Reference of Our References

Protagoras once said "Man is the measure of all things" [1]. The human being has five senses and with them he perceives and experiences his reality. In order to communicate properly the continual sequential rearrangements of reality, agreements have been made to realize uniformity in our communications [2]: mass (kg), length (m), time (s), electric current (A), thermodynamic temperature (K), amount of substance (mol) and luminous intensity (cd). The Data Processing and Transmission Delay (DPTD) between awareness and event is for the human body about 200 milliseconds. (1/5 sec) [3].

Research for the limits of our universe brought new scales: The Planck Length and Time (1.6×10^{-35} m and 5.39×10^{-44} s) were used as the lower limits of our measurable universe and the light-year (distance travelled by light (c=299,792,458 m/s) in one Julian year) as the new reference in the universal scale. Comparing the age of our universe (as for now 13,5 billion years) with the Planck time gives about 2^{200} Planck times, a number beyond understanding. If we take the Planck length scale to be 1 meter then an atom would be as large as the whole (visible) universe (93 billion light years). In this way it becomes understandable how easy neutrino's pass easily through our bodies and the earth, how much "emptiness" there is and how much the lower

---

[*] Correspondence: Wilhelmus de Wilde. E-Mail: wilhelmus.d@orange.fr Note: This article is based on a submission to FQXi Essay Contest "Questioning the Foundations, Which of Our Basic Physical Assumptions are Wrong?" in 2012.

and upper scales differ from each other. The relative human scale becomes immense and the "wave" characters the inability of perception. The reductionist way of thinking does no longer apply.

In our reductionist way of thinking, going to the essence of these references means to try to find the reference of the references (ad infinitum) and then becoming aware of the fact that we always fall back on our indescribable awareness of an emerging reality that cannot be caught.

On the experience of time, it is emergent from our memory so...the snake is going to bite his own tail: ouroboros.

Time

Time in our daily lives means the past, the present and the future. It seems easy. The past has gone, so no longer exists, the present has already gone when we become aware of it, so it is also already the past the moment that we are aware of it, the future is not yet existing, just simple. Time does not exist. It only exists as an emerging memory experience in our consciousness.

Every clock you are using/observing is at a certain distance to your senses, a distance in time (length divided by the speed of light c is the time needed for the signal of the clock to reach you), after that it takes your brain and mind still 200 milliseconds (The Data Process Transmission Delay = DPTD) to become aware of what you perceived, so we are only becoming aware of the past. Clocks measure "time" in comparing frequencies, the ultimate frequency is the Planck time (see page 1), the rhythm of the music forming our reality of this universe is $5,39 \times 10^{44}$ ticks per second which also means that there are 5.39×10^{44} possible choices per second at this rhythm.

What we are conscious of is not this frequency, we are aware of a continuous time experience, that is formed in our mind out of the pieces of mosaic of our memory, together forming a "picture" of our reality. When listening to music the treatment of the past is becoming visible, our mind is uniting separate notes as heard, the melody created in the consciousness of its composer is transmitted to us, we even form a future "expectation" so that harmony is created. One of the abilities of our consciousness is to create an image of the future, inferred of the signals received from the past.

Our senses are the recipients of the signals we receive from the outside world. Our eyes form only a "sharp" image (full colour and fully detailed) representing a surface of only 5% (the surface of a dime held at an instance of 1 meter from the eye) of the total image that we receive, the rest is **added and filled in by our consciousness**. In our minds the progress of time and so history is "created" by using "only" 5% of these data received and stocked up. This almost deniable amount of information together is forming our "subjective reality".

Subjective Reality

An individual being can be regarded as the centre of reception of signals, the OBSERVER. Around him/her is a sphere of incoming data. We could imagine of course an infinity of spheres around the observer, each sphere with a different radius and representing a different past of incoming simultaneous signals. The minimum radius of a sphere around the consciousness of the observer, at which the data are "aware" is at a distance of 1/5 sec. (10^{43} Planck lengths!). This distance represents the minimum Data Processing Transmission Delay (DPTD) between the reception of the incoming signals via our senses and the conscious causal awareness of the event observed.

Consciousness is an emergent (from a specific material cohesion) and a non-material entity that has no comparable volume. It can be represented as a singularity (no material dimensions). This singularity is the centre of the sphere. The resolution of this sphere is comparable with a digital camera, but with an almost infinite number of pixels. The deeper you go the more details are revealed. New pixels are giving new interpretations. The further we research, with the aid of instruments, the more details we will perceive. Like the digital camera we can only treat the signals that we are built for, and, as we are perceiving only 5% (baryonic matter) of the total of signals, it means that perhaps with other "senses" we will be able to get a more complete perception of our reality.

Imagine the signals from a far away past as deep blue and the younger signals as deep red, the sphere is then comparable with a multi-colour soap bubble .The surface of each sphere represents a simultaneity of data, **the subjective reality** of the observer. It contains all the history of the universe, but each percipient cannot reach all of this information, by lack of instruments and/or interest. This "knowing" (wave collapse) becomes the origin of the decoherence when shared with other subjective simultaneity spheres.

The signals arriving at the surface of this sphere come from different sources and from different distances. The sphere from observer Alice has a different (coloured) surface as the sphere from observer Bob, who is at distance x from Alice. So each observer is aware of a different reality. The minimum subjective awareness radius of 10^{43} Planck (200ms) units can be extended to other bigger radius (in length/time), until all the spheres touch and mutual cutting circles are emerging. The subjective consciousness is then more and more influenced by the mutual received signals which is the origin of decoherence, every percipient can, if he wants to, dive into this multiple conscience, but he can also decide "not to be interested", which is a form of free will.

This difference in experience of reality lies in the difference of interpretation of the subjective simultaneity spheres. If two observers communicate with each other they share a similar interest "sphere" , it will be like two soap bubbles that merge. A great communal circle where the bubbles melt together is the result. This circumference of the cutting circle is where the incoming data from observer Alice and Bob are the same. So if two observers are in different light cones in space/time they may have, dependent on their distance and communication possibilities, no communal circle, so no simultaneous experiences. It is as if they are living in separate universes.

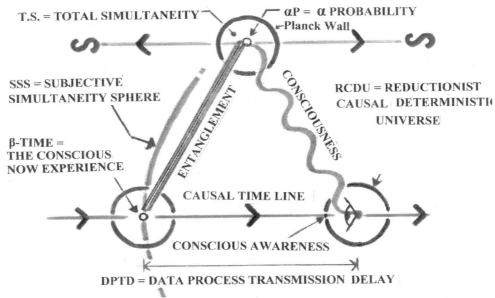

Figure 1. How consciousness is at the origin of our Now Experience. Consciousness → α-probability in TS → β-time → DPTD → conscious awareness

Consciousness Foam of Subjective Simultaneities

All the Subjective Simultaneity Spheres (SSS's) - that represent the subjective realities with their communal cutting circles - are forming together a FOAM of mutual experienced realities.

Imagine the earth with all its human beings and their foam of consciousness surrounding it, the further we are going away from earth, the further we go in the communal past and the earth is becoming more and more a centre. All of the consciousness of humanity on earth forming a new centre of the foam of human consciousness. This is the sphere of simultaneity of history around us. That is why we all share the same history, the Objective Simultaneity.

So the oldest communal image of our history is the image of the Cosmic Background Radiation, (George Smoot (Nobel Prize Physics 2006) commented the image of the CBR on the 23rd of April 1992 : "If you are religious it is like looking at the face of GOD").

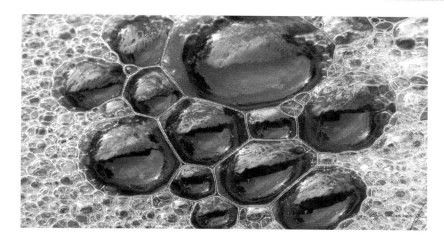

Absolute Simultaneity

The above described foam represents the subjective simultaneities and its resulting different objective simultaneities. In physics however there is also the expression: "absolute" simultaneity.

"Absolute Simultaneity refers to the experimental establishment of coincidence of two of more events in time at different locations in space in a manner agreed upon by all observers" [4].

As indicated above objective simultaneity is dependent on the specific moment in time-line (radius of the sphere) and the communal cutting circles of these individual spheres at that time. This specific moment can be described as a slit from "Block Universe". Space is not needed here as a material medium, our memories are timeless and space-less, they are the result of the data perceived on the simultaneity sphere. The absolute ether that Einstein introduced later to explain Newton is a material medium needed to explain material absolute simultaneity. In our perception this ether is of a different quality because consciousness is the origin of matter, the ether connecting "entanglement" as a spooky action at a distance is of a non-comprehensible entity for us mortal beings. It is like an imaginary ether where the cutting circles of our SSS's are, like the "distances" in time. Time is emerging only in our consciousness. The ether of consciousness is our BELIEF.

I agree with Julian Barbour when he posts : "An instant of time is ONE configuration of the entire universe at One instant" [5] so time does not exist as a dimension. The "One Instant" in our perception is the Planck time[6], the shortest time lapse until now. In this time lapse there is no longer a cause before an event. The according length is the Planck length and is derived from the Planck time. "Space" is an "emerging" perception from Time, an illusion that is created in our consciousness. If we try to explain space with reductionism , infinities and paradoxes will be all over.

α - Probability

The question can be: "Was consciousness created by matter or matter created by consciousness?" This question seems a typical chicken and egg one posed in our reductionist causal deterministic universe (RCDU), where we cannot deal with infinities.

Once we make a division line between this Reductionist Causal Deterministic Universe (RCDU) and the dimension where causality is not "existing", the one behind the until now accepted Planck length and time, where there is no cause and event relation, where we enter a "dimension" where every probability (also of parallel and multi-verse) is a possible reality. We call this "dimension" Total Simultaneity "[7] (TS). We have created the limit of our RCDU and annihilated all infinities and paradoxes of length and time. So before the Planck length we enter non causal TS, here the 3 directions that we call space do not exist nor do before nor after, we enter a ZERO-dimension Universe. The limit of our RCDU can be reached **at any point in our space** just by dividing any length until the Planck length limit (not only by going in the past!) **everywhere** we reach the origin of our universe, because it is in our consciousness.

The origin of our RCDU which is TS might be described as a singularity everywhere present. In time we can go back in the past of course in the direction of what is accepted as the Big Bang, but we can also divide every RCDU moment until its most little unit: the Planck time, this moment is the "**eternal non causal singularity moment**" and contains the whole **past and future** as experienced in a RCDU moment. The Big Bang in this interpretation is a non existing moment in our RCDU because it is a singularity, beyond the Planck time in TS. TS is "constituted" of α – **Probability's** (α-P's).

The Hawking/Hartle's No-boundary Proposal

"The Big Bang is no longer the beginning of time in a singularity as proposed in the standard inflationary cosmology **but a timeless point where the universe -or rather a superposition of all possible universes -** pops into existence from nothing with all its laws of physics intact". [8]

This touches very much the TS interpretation. However the Hawking/Hartle interpretation places the lines of all possible histories of our RCDU after the point of creation the Big Bang a timeless point in our RCDU. Our interpretation puts this point before the limit of the RCDU at a point α-P. in TS. **Therefore all the histories of our RCDU's are everywhere in TS and an eternal process.** The non causal α- P singularity in Total Simultaneity harbours ALL histories of ALL the possible universes. The "no-boundary" proposes that the "histories" of all the possible universes are positioned in the past of causal space/time, the no-boundary proposal implies no boundary in the far past but gives a boundary in the future. TS gives no-boundaries in the past nor in the future. The way however that Hawking/Hartle treat the histories and the cosmological constant Λ with the Wheeler de Witt Equation [9] may also be a mathematical working method to describe the probabilities and world-lines in TS.

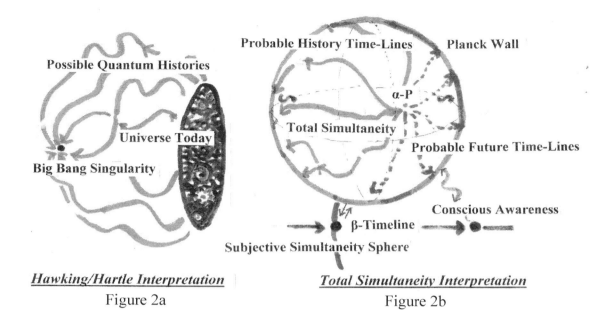

Hawking/Hartle Interpretation
Figure 2a

Total Simultaneity Interpretation
Figure 2b

Reality and Consciousness

Let us imagine an α- P in TS, this non causal probability point that is called "Alpha". This Alpha singularity is like a slit of a Block universe as the infinite oneness with no limits and no content by means of materiality of number. The α- P has a connection with our consciousness in the way that the causal part of consciousness acts like an antenna receiving/emitting timeless waves, resulting in a causal "Beta" time in RCDU. (see figure 1) so makes emerge this Beta time in our consciousness.

When we are making an observation that causes the collapse of a wave function, this observation is becoming an event and afterwards (min. 200ms) "conscious" observation. Our mind is receiving the signals from this event in β-time. The signals have to become awareness (knowledge). The causal time needed for our body (material intermediary) is about 200 milliseconds. So every observation that is the cause of a wave collapse, **is instigated in the future by our consciousness!** It took place in Beta time **before conscious awareness**. What does this mean?

The Alpha time is non causal and "existing" in TS, available to our consciousness. The beta time is causal (Arrow of time) and existing in RCDU. Our consciousness is able to make contact with the non causal Alpha time, in this way "creating" the causal Beta time. The Alpha and Beta time are in a non causal way simultaneous, the Alpha time is an eternal timeless probability moment, our consciousness is placing these moments in a Beta sequence (experiencing the arrow of time) that is causal so "understandable". It "seems" for our experience as if the Alpha moment and the resulting Beta moment are both in the past of awareness. The Alpha and the Beta moment are entangled, intelligent observing seems like creating the beta moment in our world line in the past, and **our consciousness is acting in the future (** relative to the causal arrow of time).

The double slit experiment and α-P

The "double-slit" experiment, sometimes called "Young's Experiment" is a demonstration that matter and energy can display characteristics of both waves and particles. In the basic version of the experiment, a coherent light source such as a laser beam illuminates a thin plate pierced by two parallel slits, and the light passing through the slits is observed on a screen behind the plate. The wave nature of light causes the light waves passing through the two slits to interfere, producing bright and dark bands on the screen. (Wikipedia, the free encyclopaedia: Double Slit Experiment)

By detecting particles at the SLITS, it "seems" that we have changed the result at the black wall behind. Our observation has caused TO BE particles and to stay particles.

What if we arrange the experiment so that we can make an "arbitrary choice" [10] later, whether or not to use the information gathered at the slits? When mixing up the data with irrelevant garbage data and record the combined **incomprehensible** data, the result will be "wave". When designing a program to filter the garbage out it will be "particles".

The Beta time where our choice for measurement is taking place is causal. In Alpha time it is not it is timeless. The so called choice moment is already an (ever) existing one in TS. It seems very strange in our RCDU but is "understandable" with our triangle Consciousness→ α-P → Beta. The time lapse of 200 milliseconds necessary to create awareness is not the only possible difference in causal Beta time, the definitive reality is created by our KNOWING, or our choice to have information available and manageable, which is actually taking place in future, in formula:

Conscious Awareness of measurement (particle)→α-P → = β-time + >200millisec

Wheelers "delayed choice" experiment, proposed by John Archibald Wheeler in 1978, is a variation on the double slit experiment, he directed the light into two more remote telescopes, each one focussed on one of the slits (so after the slits). By using instead of two telescopes focussed to either side of a black hole, the delayed choice is becoming light-years. In 2007 the first "clean" experimental test of Wheelers ideas was performed in France by the team of Alan Aspect et al., and it became clear that some photons must have reached earth via EACH pathway. So our observations are influencing the far away past, which means in my perception : actual conscious intervening causes a certain α-probability to become a point on the β time-line.

Spooky Action at a Distance

From entangled particles becomes also understandable. The causal moment that we are observing one particle the other will give the same outcome. Even if it is light-years away , the "change" is immediate. Our observation/measurement is taking place at causal time Beta, that moment has a non causal timeless counterpoint in TS: α-P : , so there "exists" an eternal simultaneity in time and place for every causal moment of our Universe in TS. Alice is measuring the spin of an entangled electron at A and the electron with BOB gets the same spin, no matter the distance between Alice and Bob , the reaction is timeless, the world-slice in α-P gives a simultaneity point on the cutting circles of their Subjective Simultaneity Spheres :

Conscious Awareness of Entanglement:
Simultaneity point of SSS of Alice and Bob→
Conscious Awareness of measurement (spin A/B)→α-P(spin A/B)→β-time+ >200millisec

Many Worlds Interpretation (MWI) [11]

"The wave-function collapse led to the many-worlds interpretation. The moment a wave system is observed, the wave-function collapses and the universe is split in two. The observer and the observed have become correlated or entangled"

The MWI poses that the conservation of energy is not violated at the continuous "splitting" of the universe, (huge masses and energy are needed at each splitting) because of the fact that the new universe is a universe on its own (with its own histories) where the conservation of energy is not violated. But how is conservation of energy in the total Multi-verse? The whole problem of conservation of energy (in the past and in the future) is solved when we take a closer look at our Total Simultaneity interpretation:

All the possible "past and future choices/observations/measurements points" (α- P's) of all the probable and not probable universes are "present" as α- P's in TS, it is not an act of "creating" a new material universe, but changing to another probable world-line in TS. The consciousness of the observer can go at each α- P in different ways, his consciousness develops itself further on a new world-line, the centre of his consciousness receives a new colour point on his Subjective Simultaneity Sphere (SSS). The other researchers (on the new world-line) participate in their communal circles at the merging of their Subjective Simultaneity Spheres. Once time evolves and communication is taking place, more communal circles cross and the foam is "coloured" by the new information. This growing of "knowledge" can be compared with **decoherence**, like decoherence it can be isolated (information blockage).

So the understanding "Many Worlds" gets a new interpretation, the many worlds become "many probabilities". The slice in α- P represents for the consciousness of the observer his whole world-line. All other world-lines already exist as non aligned α- P's in TS for the observer, again it is the percipient that aligns these α- P's into memory world-lines in his consciousness.

Time Travel

The splitting in the original MWI theory goes only forward in time, not backwards. In our conception it IS possible that our consciousness "activates" α- P's that were already α- P 's in our causal past memories (or in parallel universes) or were part of memory points of other percipients . Should this mean that time travel is possible? Yes but...

1. Until now our consciousness is not able to **intentionally** manage the choice of α-P's.

2. An observer may incidentally "activate" an earlier used α-P, so revive (as present) a part of the past, once that this happened his world line is split in two (in TS), one world line where he chooses to take the line from the past and one where he chooses to stay on the "old" world line, and he continues with a "memory" of another α-P in a new world line that seems to be the past of the former, but with the memory of the first. (This is also valid for so called "future" moments (α-P's from future β-time lines), one of the results of this kind of time-travel is known as precognition.

3. The grandfather son paradox does not exist because of the fact that in the new world line the grandfather is not the causal grandfather of the subject, the causal grandfather is or was in the former world line.

One way of humanity to "control" its consciousness and influence the "activation" of α-P's is perhaps meditation and praying. The most important will be the further development of our brain and/or the development of our social intelligence (objective consciousness), mankind is just at the beginning of its evolution, so there is hope.

Another way is to develop a new form of consciousness that has other sensibilities in TS, compared to our own five senses but can communicate with our consciousness. The possibilities of quantum computers indicate that this kind of "creation" could be possible in the future after all it IS a probable world-line in TS.

Entropy

Entropy is the thermodynamic property toward equilibrium/average/homogenization* Entropy is a pure causal interpretation of the past and is defined phenomenological by the second law of thermodynamics. Entropy is an expression of disorder and/or randomness, the higher the entropy the higher the disorder. TS is compared to our RCDU chaos, the highest entropy.

But is the "order" of our memories the ultimate order? Our memory gives only ONE unique world-line, with its own relative unique order. When we cut all the words out of a dictionary and throw them up, the result seems chaotic because we compare it to the former "order". The new order however is as UNIQUE as the first one. Once we throw them up again, a new unique order and so on ad infinitum. Each new point in the world-line contacted by our consciousness in TS is a creation of new order from chaos. What we perceive as chaos is in fact another form of order.

Free Will

Free will can be easily understood and accepted in this interpretation, because it is our consciousness that chooses between an infinity of α-P's in TS. Each and every moment we make these choices (of course after interpretation of our Subjective Simultaneity Sphere). The "colours" on this sphere are influencing our definitive decisions. But our decisions are not pre-dictated in a reductionist way because it depends of the "sensibility" of our consciousness how the future α-P's are chosen, EVERY α-P IS **AVAILABLE**.

In the beginning of the essay we mentioned the 200 milliseconds as the Data Processing Delay (DPL) between the reception of the signals of an event and the awareness of the ego that received them. These 200 milliseconds are 10^{40} Planck units! This means that in fact $5,39 \times 10^{40}$ choices can be made. The 200 milliseconds may seem a short time in our macro human life-time but it is like an eternity in quantum time, all is relative.

Spirituality and Its Parallels with Total Simultaneity

The Egyptians

In 3150 BC, the Egyptians had a polytheistic interpretation of their Divine ideas. What is striking in our perception is however their idea of the creation and functioning of the world that was the act of NOUN, the dark and CHAOTIC primordial ocean, that was the origin of an island ATOUM, representing the earth, or what they perceived as the universe. Here we meet the "order from chaos" as is expressed (in a different way) the order from the chaos of Total Simultaneity. The island of our "reality" is created by our consciousness.

The Greek thinkers

It was first ARISTOTLE who integers in his noblest celestial spheres the ether that wholly encloses the existent universe and calls this the "UNMOVED MOVER", the cause of everything without being CAUSE. This ether is non corporal, impersonal, immobile and not accessible for the human being. Total Simultaneity has almost the same qualities: only it is not itself the cause of everything, it is our non-causal consciousness that is harbored in this dimension that together with our causal consciousness is the cause of everything. However our consciousness is the UNMOVED MOVER. (Aristotle: Physics 8.6, 258b26 – 259a9)

Christianity

The interpretation of Total Simultaneity can be regarded as a way to "imagine" the GOD experience, parallels with Christianity can be found in the Holy Trinity:

"The Father": TS (the total of all universes and their origin),
"Jesus Christ": the human being (Us) with its causal part of consciousness:
"The Holy Ghost": Consciousness creating the order out of CHAOS and is also a part of this CHAOS (GOD?).

In this way our causal consciousness is also part of the non causal part (divine?) and the Universe as we are experiencing it cannot be other as a fine-tuned one for our form of life and consciousness (thanks to the Objective Simultaneity Spheres that are the cause of decoherence). TS (GOD?) can also give us the peace for our souls because ALL the α-probabilities are "eternally existing singularities" available to our causal and non causal consciousness, so the life-lines as we are experiencing them are also "eternal" in TS. This can explain the "contact" we have with people who died, who are in "heaven". Why can a four year old boy play a Mozart piece better as a grown up pianist [28] ? Somehow he is connected to the α-probability life-line in TS of a former pianist. Also the possible future life-lines are eternally "available" for our consciousness, the reason why some people are able to become aware of the future.

Judaism

The Kabbalah "Ein Sof" is understood as GOD prior to His Self Manifestation in the production of any Spiritual Realm and can be translated as "no end" there is no end and no beginning it is infinite. It is the origin of all created existence, the Total Simultaneity, containing ALL non causal consciousness.

The Vedanta School of Indian Philosophy

Atman: Self (the "I", the consciousness centre of the SSS) is the first principle, the true self of an individual beyond identification with phenomena, beyond the realisation of the β-time, where all probabilities are still possible choices available for the consciousness. A human being must acquire self-knowledge – atma jnana – which is to realise that one's true self (atman) is identical with the transcendent "SELF" Brahman. If Atman is Brahman in a pot (the body) then one need merely to evolve out of the pot to fully realise the primordial unity of the individual soul with the plenitude of Being that is the Absolute - the unifying of the causal and non-causal consciousness in TS). Brahman is the One Supreme, universal Spirit that is the origin and support of the phenomenal universe! Brahman is sometimes referred to as the Absolute or Godhead which is the Divine Ground of All Matter, energy, space, being and everything in and beyond this universe (TS!) Brahman is conceived as both personal (causal consciousness) and impersonal (non causal consciousness) without qualities and Supreme. If you subtract the infinite from the infinite, the infinite remains!

Hermeticism

The All: According to to hermetic doctrine, The All is more complicated than simply being the sum total of the universe. It is more correct that everything in the Universe is within the MIND of the All, since the All can be looked upon as MIND itself. Non causal consciousness is part of TS, so the universe partially exists on a Mental plane. In our perception the Universe is wholly existing as a result of the mental plane. In Hermeticism you cannot say simply "I am God" you are part of God (like our causal consciousness is not the Whole but a total with the non causal part). We have the potential of the perfection of GOD but you cannot reach it.

Sufism

"All things and events perceived by the senses are interrelated and connected and are but different aspects or manifestations of the same ultimate reality. "Enlightenment" is an experience to become aware of the UNITY and Mutual Interrelation of All Things, to transcend the notion of an isolated individual self, and to identify him/herself with the ULTIMATE REALITY".

"The direct mystical experience of reality is a momentous event, in the realm of human consciousness (as-Shuhud)"

"Instead of a linear succession of instants, they experience an *infinite, timeless and yet dynamic* present. In the spiritual world there are no time divisions such as the past, the present and the future, for they have contracted themselves into a single moment of the present where life quivers in its true sense"

"The reality underlying ALL phenomena is beyond all forms and defies all description and specification, hence to be empty, formless or void. To the Sufis all phenomena in the world are nothing but the "illusory manifestation" of the MIND and have no reality of their own" [29]

The parallel with TS as a singularity that cannot be understood/described is obvious (contracted to a single moment).

Conclusions

Our consciousness in Reductionist Causal Deterministic Universe ("RCDU") emerged from a certain order of particles. The experience of time emerges from our memory. We experience our causal order as unique and solely created for the "I" which is the centre of our consciousness. The "I" has emerged from the experiences memorized. This "I" in the centre of our Subjective Simultaneity Sphere, whose surface is partly participated with other Subjective Simultaneity Spheres of other "I's", is forming together with them a bubbling surface of "creation" which is the origin of decoherence and the awareness of our history/reality. All these reality experiences are created thanks to the ability of "acting in the future" of our RCDU consciousness because of its unity with the Alpha–probability consciousness in Total Simultaneity. Our scale of reality is an emerging one and reductionism ad infinitum will not give us the origins of WHAT IS and the WHY we are looking for. Just as Richard P. Feynman said: "We just have to take what we see, and then formulate all the rest of our ideas in terms of actual experience" [27].

This approach of "Total Simultaneity" is one of the many theories (like string theory) that seek their origin beyond the Wall of Planck. I will never claim that it is the only truth, because the only TRUTH being not 100% available for us mortal causal beings. It is however based on the latest scientific knowledge, as we all know that this knowledge is always changing because each moment new enigma's can be discovered. However our assumption that the limit of causality which is the Planck time/length is not influenced by new achieved limits, the acceptation of a causality limit is the base. The structure of TS and the perception of consciousness being ONE in two universes leads to compare TS with the ancient and modern understanding of GOD.

What Total Simultaneity and the consequences for consciousness are indicating is : We try to explain that the materialistic reductionist approach is only valid until a certain limit. The physics as we are experiencing it is "emerging". What we experience as individual "reality" is a product of consciousness entanglement and decoherence is its origin of our mutual history. Solipsism is not a consequence of this vision, Anthropomorphism is an expression that can be used here only as the indication that our reality is created by our consciousness (the causal and the non-causal part together, being ONE) and not the consciousness of another intelligent race with different constants, our reality is one of the infinite availabilities (not already "existing" space/time-lines), so is by its existence the result of our consciousness entanglement and decoherence is already fine-tuned.

References

1. Protagoras, (490BC-420BC) Pre-Socratic Greek Philosopher in Dialogue.
2. International System of Units (SI). Wikipedia, the free encyclopedia.
3. New Scientist, 14 May 2011. The Grand Delusion.
4. Absolute Time and Space, Wikipedia, the free encyclopedia.
5. Barbour, Julian: http://fqxi.org/community/forum/topic/1339 . lecture from Julian Barbour, at the College Farm, South Newington UK : Space and Time 100 years after Minkowski. His contribution : Was Spacetime a Glorious Historical Accident? (pdf), : UK 2008-10-01, other speakers: Bruno Bertotti, Edward ,Anderson, Brendan Foster, Bryan Kefleher, Karl Kuchai and neil O. Murchadha.
6. Christian, Joy, Absolute being vs Relative becoming. arXiv : gr-qc/0610049v2 23 april 2007.
7. de Wilde, Wilhelmus, Realities out of Total Simultaneity. http://fqxi.org/community/forum/topic/913
8. James B. Hartle, S.W. Hawking, Thomas Hertog : Accelerated Expansion from Negative Λ arXiv : 1205.3807v2 [hep-th] 30may 2012.
9. Wheeler-DeWitt equation - Wikipedia, the free encyclopedia.
10. Mathematical Foundations of Quantum Theory, edited by A.R. Marlow, Academic Press 1978.
11. Many-Worlds Interpretation, Wikipedia, the free encyclopedia.
12. David Deutsch, Quantum Mechanics near closed time like lines , Oxford University 1991.
13. Richard P. Feynman, The Character of Physical Law (with introduction of Paul Davies),
14. George Smoot, 23 April 1992, commenting the images of the COBE satellite: "If you are religious, it is like looking at GOD".
15. John D. Barrow, The Book of Nothing. Vintage 2001. ISBN 0 09 928845 1
16. Craig Callender/Nick Hugget : Physics and Philosophy at the Planck Scale. Cambridge University Press 2001. ISBN 0 521 66445 4
17. Alan Guth, Het Uitdijend Heelal (original : The Inflationary Universe ,1997) , Uitgeverij CONTACT, Amsterdam/Antwerpen , 1998.
18. Brian Greene, L'Univers Elégant 5the Elegant Universe, Editions Robert Laffont, S.A. Paris 2000.
19. Trinh Xuan Thuan, Les Voies de la Lumière ; FAYARD, 2007.
20. Lee Smolin, Rien ne va plus en Physique (The trouble with Physics - The Rise of String Theory, the Fall of a Science and What Comes Next, 2006, DUNOD, Paris 2007 .
21. Leonard Susskind, Le Paysage Cosmique (The Cosmic Landscape), Editions Robert Lafont, S.A. Paris, 2007.
22. David Bohm/F. David Peat, La Conscience de l'Univers. Alphée 2007. ISBN 978 2 7538 0237 7
23. Etienne Klein, Discours sur l'Oribine de l'Univers. Flammarion 2010.
24. Igor and Grchka Bogdanow : Le Visage de DIEU , Editions Grasset & Fasquelle 2010.
25. Stephen Hawking/Leonard Mlodinow : Y a-t-il un Grand Architecte dans l'Univers. Odile Jacob 2011. ISBN 978-2-7381-2313-8
26. Eric Verlinde, The Myth of Gravity, FQXi Article , January 10, 2011
27. Richard P. Feynman, Lectures on Physics. "Six easy Pieces" Addison-Wesley. 1964
28. http://www.youtube.com/watch?v=omuYi2Vhgjo
29. Dr. Ibrahim B. Seyed. . President Islamic Research Foundation International, Inc. Sufism and Physics http://www.IRFI.ORG

Exploration

Self-Awareness and Memory

Narendra Katkar[*]

ABSTRACT

It is theorized that the brain has only frequency codes, carried by induced signals, including stimulations from light, sound or other senses, which travel through atomic composition of brain material and dissipate, creating tiny "gaps" or "holes" in atomic structure. These gaps or holes are assumed to be within the cellular and molecular composition in the interior of the brain. The true nature of memory is, in my view, the transformation or conversions of self-awareness signal into those frequencies of earlier received signals by passing through the infinitesimal gap in atomic structure created by said earlier signals.

Key Words: consciousness, self-awareness, memory.

Introduction

Few theoretical physicists have argued that classical physics is intrinsically incapable of explaining the holistic aspects of consciousness, but that quantum theory provides the missing aspects (Searle, 1997). However, some physicists and philosophers consider the arguments for an important role of quantum phenomena to be unconvincing. Physicist Victor Stenger (1992) characterized quantum consciousness as a "myth" having "no scientific basis" that "should take its place along with gods, unicorns and dragons."

The association of brain activity to conscious intentions was supposed to be the basis of the functional microstructure of the cerebral cortex. The nerve impulse causes the discharge of source molecules by the course of exocytosis; it was presented as a quantum mechanical model for it is based on a tunneling process of the trigger mechanism. (Schwartz, Stapp and Beauregard, 2004)

Contemporary basic physical theory differs profoundly from classic physics on the important matter of how the consciousness of human agents enters into the structure of empirical phenomena. The new principles contradict the older idea that local mechanical processes alone can account for the structure of all observed empirical data.

Several investigations and theories relating to brain function and physics were postulated as early as in 1955, 1958 and later (Bohm, Bohr). The only acceptable point of view appears to be the one that recognizes both sides of reality—the quantitative and the qualitative, the physical and the psychical—as compatible with each other and can embrace them simultaneously. (Pauli, 1955)

---

[*] Correspondence Narendra Katkar, International Research Center for Fundamental Sciences (IRCFS), India.
E-mail: narendra.katkar@gmail.com

In a complementary procedure, averaging techniques have been used to record the electrical fields generated by the brain in the willing of a movement, the promptness potential. In exquisitely designed experiments, Libet has discovered that in conscious willing has a cerebral activation about 200 ms before the movement.

From pure basic physics point of view, a reader would be interested to know that while reading this manuscript, the words on the page are only a reflection of light. In other words, the reader receives light from the page. This reflected light induces or stimulates neuron "spike" in the brain, which re-activates the previously registered audio signals, i.e. Memory. Memory is reactivation of previously registered signals which were received through neuron spikes. Since childhood and early, a word, name or description of a thing exists in Inertia in the human brain before reactivation.

Except for a new word, the searched meaning is again the reflecting light of the printed word from a Dictionary page or an audio description, which is then superimposed or juxtaposed with the new word visual. This phenomenon of brain mechanism is examined in many disciples concerning memory and perception. Normally, humans are inclined to assume that the memory functions like recording apparatus, which is a false assumption. The molecular mechanisms essential to the induction and continuance of memory are very dynamic and consist of divergent phases covering time periods from seconds to a lifetime. (Schwarzel & Mulluer 2006)

The optic nerve contains retinal ganglion cell axons and support cells, leaves the eye socket orbit through the optic canal, leading towards the optic chiasm, which is situated at the base of the brain underneath the hypothalamus (Colman, 2006). An axon usually transmit neuron signal, an electrical impulse away from the neuron's cell body or soma. Large numbers of axons of the optic nerve terminate in the lateral geniculate nucleus (LGN), which is the primary relay center for visual information received from the retina and it is situated inside the thalamus of the brain. (Goodale, & Milner, 2004).The optic radiation or the geniculostriate pathway is a set of axons from relay neurons in the lateral geniculate nucleus of the thalamus suppose to transmit visual information to the visual cortex.

The critical question in cognitive neuroscience is about encoding and representation of information and mental experiences. It is not clear how the neuronal changes implicated in more intricate examples of memory, mainly declarative memory that necessitates the storage of facts and events (Byrne 2007). Memory Encoding is assumed as a biological event beginning with perception, passing through the brain to hippocampus where all sensations are collected into one single experience. Encoding is accomplished with a blend of chemicals and electricity. Neurotransmitters are released when an electrical pulse crosses the synapse which connects nerve cells to other cells. (Mohs, 2010).

From basic physics point of view, all brain activity is of sub-atomic phenomenon, Whether an induced electrical discharge or internal self-induced electromagnetic activity, both manifest out of atomic compositions of brain matter. Fundamentally, there is no freely available signal, one of the atoms of sodium, potassium and calcium do discharge a small fraction of its own negative charge of the value of below 30 to above 50 mV. There are about 100 billion neurons in the brain, each of which forms synapses with many other neurons. The cell fires an electrical pulse called an action potential, when the potential changes considerably. The charged atoms such as sodium, potassium and calcium direct the synaptic activity (ScienceDaily, 2011).

In human brain, the memory capacity is the ability to store and recollect information and experiences. Since last century, scientists have formulated multimodal theories on Memory. Studies of memory provide interdisciplinary link between Cognitive psychology and neuroscience. Encoding of memory involves the spiking of individual neurons induced by sensory input, which persists even after the sensory input disappears. Encoding of episodic memory involves persistent changes in molecular structures that alter synaptic transmission between neurons. The persistent spiking in working memory can enhance the synaptic and cellular changes in the encoding of episodic memory (Jensen and Lisman 2005; Fransen et al. 2002)

Simple Methods & Results

Individuals from normal life (not patients) were questioned several times about their recollections of condition in deep sleep and the condition between sleep and waking state. Also several electroencephalography EEG data was analyzed which was observed, again of the normal individuals.

Repeated questioning on recollection of condition in deep sleep and before and after waking up does confirm the "self-induced" signal is indeed related to old term "ego" and I, Me, including denials as well. The self-awareness brainwave signals are active from 5Hz frequency and above and not before in 0 to 4 Hz frequencies. The self-awareness has also a "witness" function, which then allows individual to recollect and recount. In 0 to 4 Hz frequency, the individual is in Deep Sleep and never narrates that condition (Katkar, 2013)

From 0 to 12Hz to 40Hz and above appear in fully awake conditions. The self-induced data signals have content related to I and Myself, including denials as well. 'I" is "Self Awareness" though "I" is manmade audio signal within a language. The self-awareness brainwave signals are active from 5Hz frequency and above. There are 1000s of sounds in the languages spoken around the world which correspond to "I". Verily, the self-awareness signal is creation of the consciousness in the womb or before. Conversion of this into those induced signals is sensitivity to the world of information caused by receptor neurons. Above statement means that the consciousness as self-awareness signal has to convert further from 5 Hz frequency.

Since it is not possible to enter into live brain to observe the source of brain or thought activity, an uncomplicated parallel is drawn from a Movie screen mechanism. The pictures of the physical world and the characters in effect are only light rays projected on the screen. They are the light frequencies on the film frames captured during shooting. The light from the projector passes through the film frames and converts according to matrix of dots into those light frequencies which were received during shooting, these then in totality covering screen appear as images and action (Katkar, 2013)

Similarly, the data created by laser light on a Compact Disc is stored in a series of tiny dents and planes (called "pits and lands") and programmed in a spiral data track into the top of polycarbonate layer. The programmed information is read by an inbuilt infrared semiconductor laser beam of 780 nm wavelength by a lens through the bottom of the polycarbonate layer. The reflected laser beam from "pits and lands" of a CD are converted into audio visual signals of the intensities of laser beams into different frequencies corresponding the "pits" dimension and remain original when reflecting off the "lands".

It is theorized that the self-awareness signal passes through the infinitesimal gap or hole within the atomic structure. This changes the frequency of self-awareness into the frequency of the received energy, which created the gap. More precisely, it is theorized that the self-awareness frequency converts into the frequency, which correlates the dimension of the gap or hole in atomic structure. In other words, the self-signal becomes the signal of the object earlier perceived. This conversion and reversal to self makes individual believe, having memory of the object.

The normal brain function is millions of times conversion of consciousness through self-awareness, into frequencies of objects and sounds perceived. It is further theorize that when this activity is hyper and self-awareness signal is not coming back or does not reverse, the individual mental health is disturbed. Such condition of loss of self-awareness creates health and behavioral problems.

So, does the world around send any information of its own natural condition?

In the brain there is no projector, no light, no film to register external light, no screen to project the image of the physical world. Neither there is any mechanism of a compact disc for recording and reading. The image projected on the movie screen and in the brain correspond the light reflected from the bodies.

In other words, in visual perception, the reflected light from the physical world, including humans etc, may not carry any information. Indeed, it is assumed that the light after reflecting does not carry any physical, physiological, chemical, biological, molecular or atomic information of the body perceived. At the instant of impingement and reflection (in light speed) the initial frequency of light is changed, effectively, attenuates and changed frequency has the color attribute. Color and luminosity are the attributes of light. Neither there are "physical bodies" on the screen nor in the brain (Katkar, 2013).

Fundamentally, the assumed memory of physical world is, in my opinion, self-imposed "false memory". This false memory held strongly or obsessively in the brain is conflict prone and creates disturbed mental conditions. It can be inferred that this memory, only for practical reason, embedded in the day-to-day lives of individuals, helps organize life.

The memory reactivations from 5Hz up to 12 Hz appear between wake-sleep states. This is the condition where an individual is neither fully awake nor in deep sleep. The narration of images, called dream, are of different intensities hence the individual can sometimes narrate those images clearly and at other times he or she cannot recollect the images.

The above two states of dream images correspond to high and low intensities of brain frequencies. Between 8 Hz and 12 Hz of brain waves do carry certain intensity of image resolution, which then, the individual recollects and narrates. The low intensity of image resolution, which appears between 5 and 8 Hz of brain frequency, is not clearly remembered. The individual may express indistinct recollections of some images, which are obscure visuals, manifested just after deep sleep condition. In other case, the frequencies are near to waking state as the intensity is higher hence the possibility of remembrance. In a few other cases, due to higher frequency activity, between 8 and 12 Hz, individual experiences ad-mixture of visual data which creates a non-cohesive image display or dream sequence. The energetic activity corresponding induced signals by sense perception is in fact consciousness is active in energetic form. In other words, active consciousness is energy.

Discussion

Research shows that these negative charge (neuron signals) carrying the light frequency information rest in the nucleus of lateral geniculate, with the frequency codes. When the external stimuli re-activate these past codes, the brain has the faint image of that physical perception. These electromagnetic frequencies are extremely weak. Since childhood, humans are creating a self-imposed embedded program through juxtaposing descriptive audio induced (language) signals with visual light produced signal in center of brain and these reactivate as memory. These, in pure physics terms do not represent the physical world. Indeed, in my view, the physical world does not have its own means to send its own information, either in light form or audio form.

The supposed memory of physical world was tested simply by asking the individual to walk in one's own house by closed eyes, where every object is in memory held by the individual as his/her own known physical environment. The individual could not walk freely more than three steps in bedroom to bathroom or in sitting (drawing) room to kitchen or in other places. This establishes that there really is no information of physical world in the brain and it also elucidated that by open eyes, the light frequencies from each object of one's own environment invoked the previously available frequency codes, giving individual a sense of assurance of having "knowledge" of physical surrounding to move freely.

According to basic physics mentioned earlier, the initial charge emission does in fact activate or excites other atoms in immediate vicinity, which appears as a network of neural activity. Whichever may be the cell, as described in five divisions of neurons within the retina, which are photoreceptor cells, bipolar cells, ganglion cells, horizontal cells, and amacrine cells. The basic circuitry of the retina is supposed to incorporates a three-neuron chain consisting of the photoreceptor, a rod or cone, bipolar cell, and the ganglion cell and the first action potential seems to occur in the retinal ganglion cell, which is the direct path to transmit the visual information to the brain (Purves, 2008, Ramachandran, 1998) which again must be understood as a subatomic emission out of one of the atoms in the cell composition, either out of calcium atom or potassium or sodium atom.

The signal travels around 3 to 5 centimetres inside brain and terminates or dissipates in the atomic structure, creating, in my opinion, infinitesimal hole. Fortunately, the signals dissipate, otherwise they will excite billions of atoms, which in return will radiate and brain will become degenerate and burn off. In such case, Human being, after developing five senses, will not survive, even childhood.

Samples of EEG signals show distribution of electromagnetic radiation of energy emissions. The amounts of energy observed are delta waves. A delta wave produced from deep sleep called slow-wave sleep is a high amplitude brain wave with a frequency of oscillation between 0–4 hertz (Walker, 1999; Kirmizialsan, 2006) and Alpha of 8–12 Hz detected strongest neural activity in the occipital lobe during awake and relaxed condition (Cantero et al. 2003). Theta wave is of 4–8 Hz (Cantero et al. 2003), and Beta is of 13–30 Hz and Gamma waves in 30–70 to 100 Hz frequency band (Berger; Gray, 1929, Fries P 2001, Llinas, Yarom, 1986). The brain activity or Mu waves are electromagnetic oscillations in the frequency range of 8–13 Hz and appear in bursts of at 9 – 11 Hz (Oberman et al. 2005, Churchland, 2011).

It appears that all memory activation is dependent on a stimulus. A single external stimulus or even a self-induced becomes the cause of re-activation of latent memory. In fact, consciousness is self-awareness signal converting into inactive signal. Keeping self-awareness frequency cutoff from conversion into objective signals during wakeful state is the most extraordinary function which will lead to ultra or supersensory perception (if mastery achieved - author has experienced twice). The first few experiences will be perception of existence in a non-dimensional condition and other is of no-gravity state or floating state.

Conclusion

Human brain parts are inactive after death and the live brain is in my view only energetic activity. The true nature of *Memory* is theorized as the transformation or conversions of self-awareness signal into objective frequencies by passing through the infinitesimal gap in atomic structure created by earlier received signals. Above 5Hz, the self-awareness signal is transforming into objective frequencies and also having subjective function as witnessing, which declares, I see, I know etc.; and even in the negations. On the other hand, the self-awareness signal in 1 to 4Hz is subjective and not converted into objective signals. The research continues on how to keep self-awareness frequency from conversion into objective signals during wakeful state, which may lead to ultra or supersensory perception if mastery is achieved. In my view, the active human brain is an extraordinary *"Game of Energy"*.

References

Berger H; Gray, CM (1929). "Uber das Elektroenkephalogramm des Menschen". *Arch Psychiatrika Nervenkrankenheit* 87: 527–570. Doi: 10.1007/BF01797193. PMID 7605074.

Bohm, D. J. 1990 A new theory of the relationship of mind to matter. Phil. Psychol. 3, 271–286.

Bohm, D. & Hiley, D. J. 1993 The undivided universe. London: Routledge.

Bohr, N. 1958 Atomic physics and human knowledge. New York: Wiley.

Bohr, N. 1963 Essays 1958–1962 on atomic physics and human knowledge. New York: Wiley.

Byrne, J. H. (2007) Plasticity: new concepts, new challenges. In: Roediger, H. L., Dudai, Y. and Fitzpatrick S. M., eds. Science of Memory: Concepts. New York: Oxford University Press, pp. 77–82.

Cantero JL, Atienza M, Stickgold R, Kahana MJ, Madsen JR, Kocsis B (2003). "Sleep-dependent theta oscillations in the human hippocampus and neocortex". JOURNAL Neuroscience 23 (34): 10897–903. PMID 14645485. http://www.jneurosci.org/cgi/content/full/23/34/10897.

Cecie Starr (2005). Biology: Concepts and Applications. Thomson Brooks/Cole. ISBN 053446226X. http://books.google.com/?id=RtSpGV_Pl_0C&pg=PA94.

Churchland P, Braintrust, Princeton University Press, 2011, Chapter 6, Page 156

Coffey, Peter (1912). *The Science of Logic: An Inquiry Into the Principles of Accurate Thought*. Longmans. http://books.google.com/?id=j8BCAAAAIAAJ&pg=PA185&dq=%22roger+bacon%22+prism.

Cuthill, Innes C (1997). "Ultraviolet vision in birds". In Peter J.B. Slater. Advances in the Study of Behavior. 29. Oxford, England: Academic Press. p. 161. ISBN 978-0-12-004529-7.

Fransen, E., Alonso, A.A. and Hasselmo, M.E. (2002) simulations of the role of the muscarinic-activated calcium-sensitive non-specific cation current I(NCM) in entorhinal neuronal activity during delayed matching tasks. Journal of neuroscience, 22, 1081-1097.

Fries P (2001). "A mechanism for cognitive dynamics: neuronal communication through neuronal coherence". *TICS* 9: 474–480.

General discussion: Roland, P. E., Larsen, B., Lassen, N. A. & Skinh0j, E. (1980), J. Neurophysiot. 43, 118-136.

Hughes JR. (2008). Gamma, fast, and ultrafast waves of the brain: their relationships with epilepsy and behavior. Epilepsy Behav. Jul;13(1):25-31. PMID 18439878

Ian Gold (1999). "Does 40-Hz oscillation play a role in visual consciousness?" *Consciousness and Cognition* 8 (2): 186–195. doi:10.1006/ccog.1999.0399. PMID 10448001.

Jamieson, Barrie G. M. (2007). Reproductive Biology and Phylogeny of Birds. Charlottesville VA: University of Virginia. p. 128. ISBN 1578083869.

Jensen, O. and Lisman, J.E. (2005) Hippocampal sequence-encoding driven by a cortical multi-item working memory buffer. Trends in Neuroscience, 26, 696-705.

Kirmizialsan, E.; Bayraktaroglu, Z.; Gurvit, H.; Keskin, Y.; Emre, M.; Demiralp, T. (2006). "Comparative analysis of event-related potentials during Go/NoGo and CPT: Decomposition of electrophysiological markers of response inhibition and sustained attention". *Brain Research* 1104 (1): 114–128. doi:10.1016/j.brainres.2006.03.010. PMID 16824492.

Libet, B. (1990): The Principles of Design and Operation of the Brain, eds. Eccles, J. C. & Creutzfeld, 0. (Springer, Berlin), pp. 185-205 plus General Discussion, pp. 207-211.

Katkar, Narendra (2013): Science of self-awareness and memory, *International Journal of Research Studies in Psychology,* January 2013, Volume 2 Number 1, 69-77

Mohs, Richard, C. 2010 "How Human Memory Works." 08 May 2007. HowStuffWorks.com. http://health.howstuffworks.com/human-memory.htm 23 February 2010.

Oberman LM, Hubbard EM, McCleery JP, Altschuler EL, Ramachandran VS, Pineda JA. (2005) "EEG evidence for mirror neuron dysfunction in autism spectrum disorders". *Cognitive Brain Research*. Jul; 24(2):190-8

Pauli, Wolfgang, 1955, the influence of archetypal ideas on the scientific theories of Kepler. The Interpretation of nature and the psyche. London: Routledge & Kegan Paul.

Posner, M. I., Petersen, S. E., Fox, P. T. & Raichle, M. E.(1988) Science 240, 1627-1631.Deecke, L. & Lang, V. (1990) The Principles of Design and Operation of the Brain, eds. Eccles, J. C. & Creutzfeld, O.(Springer, Berlin), pp. 303-341.

Purves, D., Augustine, G.J., Fitzpatrick, D., Hall, W.C., LaMantia, A., McNamara, J.O., White, L.E. Neuroscience. Fourth edition. (2008). Sinauer Associates, Sunderland, Mass. Print.

Ramachandran, V. S. and S. Blakeslee (1998), Phantoms in the brain: Probing the mysteries of the human mind. William Morrow & Company, ISBN 0-688-15247-3. Print.

Reproducing Visible Spectra. Repairfaq.org. Retrieved on 2011-02-09.

Schmitz D.1; Gloveli T.; Behr J.; Dugladze T.and Heinemann U. (1998). "Subthreshold membrane potential oscillations in neurons of deep layers of the entorhinal cortex". Neuroscience, 85:. 999-1004

Schwartz, M. Jeffrey Henry P. Stapp and Mario Beauregard, 2004: Quantum physics in neuroscience and psychology: a neurophysical model of mind–brain interaction: Phil. Trans. R. Soc. B, doi:10.1098/rstb.2004.1598

Schwarzel. M.& Mulluer. U., (2006): "Dynamic Memory Networks", "Cellular and Molecular Life Science",

ScienceDaily, 2011: Mimicking the Brain -- In Silicon: New Computer Chip Models How Neurons Communicate With Each Other at Synapses

Searle, John (1997). *The Mystery of Consciousness*. The New York Review of Books. pp. 53–88. ISBN 978-0-940322-06-6.

Stenger, Victor, "The Myth of Quantum Consciousness", *The Humanist* Vol 53 No 3 (May–June 1992) pp. 13-15 [1]

Thomas J. Bruno, Paris D. N. Svoronos. CRC Handbook of Fundamental Spectroscopic Correlation Charts. CRC Press, 2005.

Walker, Peter (1999). *Chambers dictionary of science and technology*. Edinburgh: Chambers. p. 312. ISBN 0-550-14110-3.

Exploration

Consciousness Dimensions & Quantum Non-locality

Cebrail H. Oktar[*]

Department of History Science, History of Science Society, 440 Geddes Hall, University of Notre Dame, Notre Dame, IN 46556

ABSTRACT

In this paper, the author reviews Carl Jung's consciousness dimensions and discusses these dimensions in the context of quantum non-locality. Jung describes four basic psychic functions for consciousness which they are intuition, sensation, feeling, and thinking. Feeling and thinking form a polarity, just as intuition and sensation do. According to Jung, people are equal to the extent that every person possesses these four basic functions. Differences between individuals are based on the fact that the influence of each basic type is of a different strength within each person.

Key Words: consciousness, dimension, quantum, non-locality, Carl Jung, intuition, sensation, feeling, thinking.

1. Introduction

Jung [1,2,3,4,5] says that "under sensation I include all perceptions by means of the sense organs; by thinking, I mean the function of intellectual cognition and the forming of logical conclusions; feeling is a function of subjective evaluation; intuition I take as perception by way of the unconscious, or perception of unconscious events."

Jung goes on to explain that, in his experience, there are only four basic functions, a fact that seems to be self-evident if one inquires into the possibilities. These psychic functions are the methods employed by humans to acquire knowledge of themselves and the surrounding world; cognition is not restricted to one function, and each function provides its own kind of knowledge. Jung's typologies [6,7,8,9,10] are the attitude types of introversion and extraversion, which he describes:

"The introvert's attitude is an abstracting one . . . he is always intent on withdrawing libido from the object, as though he had to prevent the object from gaining power over him. The extravert, on the contrary, has a positive relation to the object. He affirms its importance to such an extent that his subjective attitude is constantly related to and oriented by the object. These brief explications of his major topics, namely, the eight variations of personality and the attitude types of introversion and extraversion, are later described as having this purpose:

---

*Correspondence: Cebrail H. Oktar, Department of History Science, History of Science Society, 440 Geddes Hall, University of Notre Dame, Notre Dame, IN 46556. E-mail: javaquark@gmail.com

To provide a critical psychology this will make a methodical investigation and presentation of the empirical material possible. First, and foremost, it is a critical tool for the research worker, who needs definite points of view and guidelines if he is to reduce the chaotic profusion of individual experiences to any kind of order."

Jung said of his typology [9,10,11,12] "It is not a physiognomy and not an anthropological system, but a critical psychology dealing with the organization and delimitation of psychic processes that can be shown to be typical.

" Here Jung makes it clear that he was not concerned with the origins of the psychological functions, but used them as a tool in organizing empirical material. It was Jung's purpose to describe individual types of the human personality, to explain and explore individual differences of cognition and various methods of expression in the personality by using the psychic functions of intuition, sensation, feeling, and thinking, along with the attitudinal types of introversion and extraversion [1,2,3,4,5].

Jung states [12,13,14,15]: "Since every man, as a relatively stable being possesses all the basic psychological functions, it would be a psychological necessity with a view to perfect adaptation that he should also employ them in equal measure" (p. 19). Here Jung confirms the possibility of all four functions working in equal measure in the psyche of one person. Throughout his writing, he describes what happens when one function is superior and conscious and another function is inferior and unconscious. When one conscious position is extreme, the position of the other extreme will exist in the unconscious, causing a neurosis or a maladaptation to consciousness.

The interplay of conscious and unconscious opposites, as well as opposites in general, is prevalent in Jung's thinking and in his writing, and appears to be the foundation for his theory of opposites or the transcendent function. Jung [15,16,17,18] describes this as follows:

"The function being here understood not as a basic function but as a complex function made up of other functions, and "transcendent" not as denoting a metaphysical quality but merely the fact that this function facilitates a transition from one state to another. The raw material shaped by thesis and antithesis, and in the shaping of which the opposites are united, is the living symbol."

This definition describes the importance that Jung gave to the symbol as a means for uniting the opposites, and also describes the complex relationship of the symbol with the four psychological functions. An expanded individual consciousness was not seen as important *only* to the person who obtains the limits of personal potential, but equally important to the society to which he belongs. Jung makes this clear when he says that "development of individuality is simultaneously a development of society. Suppression of individuality through the predominance of collective ideals and organizations is a moral defeat for society"

Fordham [18,19,20,21,22] says that "The subtitle of the first English translation of Psychological types reads or The psychology of individuation, an addition to the first Swiss edition and not added to later ones in German; it is omitted from the recent edition in the Collected works. The addition is curious because there is no mention of individuation in the text until its definition at the end of the book."

This is literally true, but not quite reflective of the spirit of the text, which I understand as significantly related to the individuation process. Meier [19,20,21,22], however, appears to share my conviction concerning typology:

"Jung's most important contribution to psychology was the discovery and practice of the process of individuation. In spite of the supremacy of this concept, its origins in terms of chronology are far from being as clear as they should be, and in this respect not even Jung's memoirs yield the needed biographical information. But before I give you the prehistory, history and an account of the survival of the concept of individuation, I shall ask you to remember: Individuation begins and ends with typology."

Jung describes and links his work on psychological functions with the concept of individuation in an important way: "The concept of individuation plays a large role in our psychology. In general, it is the process by which individual beings are formed and differentiated; in particular, it is the development of the psychological individual as a being distinct from the general, collective psychology. Individuation, therefore, is a process of differentiation having for its goal the development of the individual personality."

The above definition succinctly describes Jung's purpose in attempting to provide a theoretical model of psychological types or functions. Individuation appears to me to be the primary goal of this work and Jung's multitudinous insights are, as he described them, "critical tools" for further research [6,7,8,9,10,11,12,13,14,15,16].

2. Consciousness Dimensions

Thinking

Thinking refers to the faculty of rational analysis; of understanding and responding to things through the intellect, the "head" so to speak. Thinking means connecting ideas in order to arrive at a general understanding. The Thinking-type often appears detached and unemotional. The Scientist and the Philosopher are examples of the "thinking type", which is found more commonly in men.

Feeling

Feeling is the interpretation of things at a value- level, a "heart"-level rather than a "head"-level. Feeling evaluates, it accepts or rejects an idea on the basis of whether it is pleasant or unpleasant. According to Jung this is the emotional personality type, and occurs more frequently in women.

Thinking and Feeling are both rational, in that they both require an act of Judgment. Sensation and Intuition are both irrational, in that they involve no reason, but simply result from stimuli (whether external or internal) acting upon the individual.

Sensation

Sensation means conscious perception through the sense-organs. The Sensation personality-type relates to physical stimuli. But there is a difference according to whether the person is an introvert or an extrovert. We could have an Introverted-Sensation type, such as an artist, who experiences the physical world (sensation) from the perspective of the psychic or inner consciousness (introversion). As opposed to this, the Extroverted-Sensation type would be the person who is a simple materialist or hedonist, interested only in physical or pragmatic things. This type tends to be realistic and practical. At worst, one may be crudely sensual. This personality-type occurs more often in men

Intuition

Intuition is like sensation in that it is an experience which is immediately given to consciousness rather than arising through mental activity (e.g. thinking or feeling). But it differs in that it has no physical cause. It constitutes an intuition or hunch, a "gut"-level feeling, or an experience. It is the source of inspiration, creativity, novel ideas, etc. According to Jung, the Intuitive type jumps from image, is interested in a while, but soon loses interest.

With the four Ego-faculties of Thinking, Feeling, Sensation, and Intuition we have a basic classification of modes of consciousness; one that has been postulated under various forms of which Jung's is only the most recent for centuries [17,18,19,20,21,22].

3. Non-locality of Consciousness State

The signal travels through space and, nothing can have velocity which is faster than that of light is the inviolable dictum of Albert Einstein. With this dictum, however, the classical mechanics was extended to the level of Relativity. Meanwhile, the quantum principles were also silently insinuating in the field of physics. Einstein, being terribly uncomfortable with the element of uncertainty in quantum mechanics, devised a thought experiment along with Podolsky and Rosen. The result of this thought experiment proved to be paradoxical.

This is known as EPR paradox [23,24,25,26,27,28]. It was observed that there were occasions when Einstein's own dictum was possibly not valid. Then the scientists also came across Bell's theorem [23,24,25,26] and Non-locality [23,24,25,26]. Alan Aspect from France set up an experiment [?] which unequivocally proved the existence of nonlocal communication between particles 12 meters apart in less than a billionth of a second i.e., 20 times faster than the supposedly unbreakable speed of light. The idea of non-local communication is further reinforced by the experiment of Nicolas Gisin who in 1997 demonstrated nonlocal communication over 10 km of distance with a speed 20,000 times faster than the speed of light. Nonlocal communication happens across galaxies between once related quantum particles [23,24,25,26,27,28,30].

Local communication is through space and the signal does not travel faster than light. Nonlocal communication is a kind of communication between quantum particles which have had interaction with each other in the past. The communication is unmediated by any known form of energy. The communication is also unmitigated, where the strength of interaction is independent of magnitude of separation. Nonlocal communication happens independent of space (spatial non-locality) and often independent of time (temporal non-locality). Spatial non-locality is when two events happen simultaneously irrespective of their spatial separation.

The events are causally connected, meaning the event in the place one initiates the event in the second place or vice versa. It is simultaneous happening of two causally connected events separated by space. It is instantaneous action at-a-distance. Temporal non-locality, on the other hand, dissolves the barrier of time. This communication works independent of time. The happening of two events, causally connected, irrespective of barrier of time indicates temporal non-locality. For example, a distant recipient receives a mental message long before it is mentally sent. According to many scientists, an element of time-reversibility plays here.

The scientific progress is evidence-based. The scientists have objective and reproducible evidence in their hands to accept the phenomenon of spatial and temporal non-locality in science. Spatial non-locality is also called non-locality type I and, temporal non-locality is non-locality type II. It is logical to extend our thought that if communication is possible independent of space or independent of time, it is also likely, therefore, that a kind of communication would be there which is independent of both space and time. This is non-locality type III. This is a prediction. There is no 'scientific' evidence yet obtained for this. Many scientists discard this possibility and prefer to debunk it because of its teleological flavor. However, if science ever accepts seriously teleology as a scientific issue, non-locality type III could be its immediate tool to explore it further. The issue is not so easily dismissed by the accomplished mystics. They claim to experience nonlocal communication type III quite regularly.

The author prefers to call it, 'purposal' non-locality, where the vital information regarding the goal the master plan is communicated to a system where it continues to do work independent of both space and time. Rather, it changes the earlier space-time wrap of the system to a degree that it becomes conducive for the newly set purpose/goal/master plan. This type of communication is proposed to exist amongst evolving and/or living systems. It is rare to find it in the non-evolving or non-living system.

The ontological relationship of space, time and 'purpose' is reflected in the ontological relationship of type I, type II and type III non-locality. The interesting feature is that acquisition of non-locality type II includes acquisition of non-locality type I and acquisition of non-locality type III includes acquisition of other two as well.

It is David Bohm who tried to bring an ontological distinction amongst different types of non-locality. According to him, quantum non-locality is not instantaneous. It involves some kind of signal transmission, may be at superluminal speed. However, non-locality which is instantaneous does exist which he preferred to call supernon-locality. Bohm also used the term super-supernon-locality which probably may be what we are calling non-locality type II.

There are three important disciplines in present science, each of which has grown under the respective umbrella of an inviolable constant set up by its pioneer. For the discipline of Relativity it is velocity of light, for Quantum Mechanics, it is Planck's constant and for Cybernetics it is entropy barrier. These three constants set the limit, the boundary of present science, which is based on measurability. Einstein's constant i.e. velocity of light excludes the possibility of simultaneity of events. The phenomenon of spatial non-locality, however, dissolves this barrier of space and indicates existence of events beyond the boundary drawn by Einstein. Planck's constant excludes the possibility of continuity of events. 'Discontinuity' is one of the cardinal characteristics of quantum world. As existence of spatial non-locality throws a serious challenge for Einstein's constant, the phenomenon of temporal non-locality seems uncomfortable for Planck's constant.

For many quantum mechanists, temporal non-locality is a 'paradox', almost similar to the paradoxical situation the classical and relativistic physicists faced with advent of spatial non-locality, EPR paradox. Beside Quantum Mechanics and the Relativity, the third important discipline of science is the discipline of cybernetics, which works under the inviolable umbrella of entropy barrier. This entropy barrier excludes the possibility of identity of events. The phenomenon of non-locality type III, if it exists, proclaims identity of events. The identity of events is independent of both space and time and questions inviolability of the entropy barrier in the discipline of cybernetic in its present form. The phenomenon of nonlocal communication has, therefore, a brighter side to endure with. It bears the potential to extend the present science to a deeper plane of nature and consciousness. Nonlocal communication so far has been acknowledged as phenomenon. The scientists still do not have any control over it. Therefore, they cannot use it also. The stumbling block in investigating non-locality is the non-specificity of this phenomenon in the realm of measurement. Many scientists, therefore, consider it as a 'Permissive' theory and not a 'Specific' theory. However, the future of information technology depends on how the scientists can gain access and then control over this phenomenon and use it for communication purpose. Physicist Henry Stapp is probably right when he says that non-locality is the greatest discovery of modern science.

4. Transformation of Consciousness State

Einstein once said that telepathy had much to do with physics than with the psychology[31,32,33,34,35]. We are not sure whether he was talking of classical, relativistic physics or of the relevance of quantum physics. It is proposed that the nonlocal communication could be the right choice in this context. The idea that human brain has 'biologized' quantum mechanical principles has been there for over three decades, mostly amongst the physicists. However, it is E. H. Walker who in 1970 published a paper which clearly stated the possibility of quantum nature of the brain, particularly at the level of synaptic communication [31,32,33,34,35]. Physicists like H. P. Stapp, Amit Goswami have stressed on this point [35,36,37,38,39]. Goswami also conceptualizes that 'Self' could behave in 'Classical' or 'Quantum' way, the later may be called quantum self. Nobel neuroscientist Sir John C. Eccles embraced quantum mechanics in his study of synaptic transmission [35,36,37,38,39]. Recently, Roger Penrose and Stuart Hameroff published a paper in the Journal of Consciousness Studies, relating subcellular quantum events in the neuron with development of conscious awareness [35,36,37,38,39].

There is another view that quantum streams of events within the brain are in communion with Quantum Sea outside to generate an experience of supracortical consciousness. However, it is nonlocal behavior of the brain, or nonlocal communication by the neurons in the brain which are relevant to the message of this study.

Thinking state transformation

Non-locality can never be simulated in a classical system. For nonlocal communication to occur the objects must be of quantum nature. That the brain as a whole can behave as macroquantum object and can participate in nonlocal communication has been demonstrated by Grinberg's experiment [40,41,42,43,44,45,46,47].

"Einstein-Podolsky-Rosen (EPR) correlations between human brains are studied to verify if the brain has a macroscopic quantum component. Pairs of subjects were allowed to interact and were then separated inside semisilent Faraday chambers 14.5 m apart when their EEG activity was registered. Only one subject of each pair was stimulated by 100 flashes. When the stimulated subject showed distinct evoked potentials, the nonstimulated subject showed
"transferred potentials" similar to those evoked in the stimulated subject. Control subjects showed no such transferred potentials. The transferred potentials demonstrate brain-to-brain nonlocal EPR correlation between brains, supporting the brain's quantum nature at the macrolevel."

The interacting pairs of brain in Grinberg's experiment were chosen from those persons who meditated together and could feel the presence of other partner irrespective of their separation at distance. Meditation although has stimulated a lot of research interest in the West, both for the disciplines of neuroscience and physics, the hard core scientists still hesitate to embrace it because of a lot of subjectivity inherent in the field. Non-reproducibility and closedness to public scrutiny are other reasons for scientist's repulsion to explore spiritual experience. What I suggest in the followings is that there are numerous situations, beside formal meditation, where nonlocal communication between once-interacted pair of brains remains a clear possibility. Meditation is not the only consciousness-expanding procedure. Consciousness grows as well in nurturing of love, promoting devotion and cultivating trust. With expansion of consciousness, brain starts participating in nonlocal communication. Implications of the fact that the brain can participate in nonlocal communication are far reaching.

Feeling state transformation

Firstly experimental findings on brain's nonlocal activity, as demonstrated by Grinberg et al is a serious challenge to the classical views of brain which hold brain as a wired structure to process information and to respond. Nonlocal communication is outside the scope of any classical system. It is stated to be one of the signatures of micro or macro-quantum object. For a structurally polylithic brain to behave as a macro-quantum object, it has to work functionally as a monolithic organ. How does the classical structure of the brain behave, as one ensemble, single block, a quantum at macroscopic level, is a frontier for research. Let me explain this a little. The brain has three evolutionary components nested vertically one above the other - the reptilian brain, the mammalian brain and the human brain. It has two culturally differentiated components, arranged horizontally, the left hemisphere and the right hemisphere. Although these components have been assigned different works at appropriate level, a composite global response from the brain is almost always an outcome even at the classical level of its response. When the brain participates in nonlocal communication, the question how does this polylithic structure become monolithic is likely to stimulate many scientists to revise their knowledge in simultaneous classical and quantum behaviour of some systems like brain.

Sensation state transformation

We accept the view that the brain can engage in nonlocal communication, we are able to explain many phenomena, hitherto left as anomalous. There are sensory perceptions which are not communicated through known sensory routes (extrasensory perceptions). There are also perceptions which are not 'sensory' in nature (nonsensory perception). Nonlocal communication can come in rescue to substantiate these phenomena.

Intuition state transformation

Not every brain has the ability to acquire the capacity of nonlocal communication in an integrated way. However some brains can acquire it when they successfully overcome various uncertain situations. The issue of development of integrity is related to the evolutionary potential of brain. Integration, in this context, could be defined as successful biologization of information relevant for overcoming different levels of uncertainty.

There is usually no perception of uncertainty in the classical plane. The uncertainty is felt between two complementary properties of an object (Heisenberg's uncertainty) when the observation and measurement are in a quantum plane. At a deeper plane of existence, a different kind of uncertainty is perceived between the conditioned properties and the very existence of the observed (e.g., uncertainty at the level of a black hole). In a still deeper plane, the perceived uncertainty is between 'Existence' and 'No existence', between the 'Presence' and the 'Absence' (e.g., uncertainty at the edge of the universe) and finally, the uncertainty oscillates between 'Nonexistence' and 'A new existence', between 'Absence' and a 'New Presence'. Does an ontological relationship exist amongst different types of uncertainties as mentioned? Although difficult to argue at this stage, I guess it exists. Stephen Hawking, in contrast to Roger Penrose, has already pointed out that the uncertainty at the level of a black hole is of a different kind, different from Heisenberg's uncertainty. It is also Hawking who mentions about the most profound uncertainty found at the edge of the universe. Does nonlocal communication occur while a brain or to say any macro-quantum object under evolution, overcome various depths of uncertainty? A very pertinent issue to attend, indeed.

The uncertainty can be reduced by input of relevant information into an informational open system. One of the defined properties of information is its ability to reduce uncertainty. The meaning of information is, however, read by the system in which it is introduced. Inclusion of the information within the system (systemization), may also be called 'biologization' of information, leads to a new integration within the system.

Time transformation

The capability of nonlocal communication is another added feature in the concept of time. Here, consciousness is a basic problem since it affects the process of observation as well as the process of measurement. Science here cannot be purely objective or positivistic. Observer's consciousness influences the epistemology. Quantum mechanists are, however, silent on two points, which are on the boundary of the quantum world; quantum discontinuity and quantum void, the 'sink' and 'fountain-head' respectively for the quantum plane. This implies a very existence of another terrain, which remains deeper to the quantum world. I label it as the terrain of elementary phenomena of which 'death' (de-conditioning of properties) and a new `life' or `re-birth' (reconditioning of properties) are inescapable elements.

The people are afraid of penetrating through this `discontinuity' because of profound degree of uncertainty attached to it. In this terrain, uncertainty is of a different kind, of a different ontological status. In the plane of quantum, uncertainty principle is applied on canonically conjugate conditioned properties of the observed. In the terrain of elementary phenomena, uncertainty for an observer-dependent reality is between conditioned properties and the very existence of the observed. Penetrating through discontinuity, while one explores a domain where existence itself has been put under question; there is all likelihood of a fusion of quantum language and metaphysical language. Discontinuity is not merely a metaphysical issue. It is a scientific issue too. A bridging language here is of utmost importance.

5. Summary & Discussion

Feeling and thinking form a polarity, just as intuition and sensation do. According to Jung, people are equal to the extent that every person possesses these four basic functions. Differences between individuals are based on the fact that the influence of each basic type is of a different strength within each person.

Jung comes to the following four basic types of persons:

The feeling type
The thinking type
The intuitive type
The sensational type

Every person can be classified according to one of these four basic types. The person's basic type determines the typical way of experiencing and valuating reality. Before going further into the theoretical aspects of the basic functions and relations between them, the types are briefly described in terms of behavior, which for example can be observed during a meeting. This way a quick understanding of the four basic types, the foundations of Jung's typology, is obtained.

The feeling type focuses himself particularly on the feelings that exist between people. He attaches importance to fine relations between the members of the assembly. It is the type that wants to keep people together. Results are of less importance.

The thinking type focuses himself particularly on substantial research results. He is the type that wants to have a firm, objective and analytic research to be done before he decides anything.

The intuitive type wants to see the subject of the meeting in a larger, more inclusive and strategic context. This type hates rash action, and is as such opposite to the next type.
The sensational type wants to actually put something into effect, and do it as fast as possible. In the end there will be time enough for things to be reflected.

Each type, seen from its own point of view, is in the right; each type represents an essential dimension of reality. Many conflicts and tensions between people can be brought back to differences in the way people look at reality and value reality.

References

[1] Jung, C. G. Modern man in search of a soul. New York: Harcourt Brace & Co., 1933.
[2] Jung, C. G. Psychological commentary on the Tibetan book of the great liberation. In Psychology and religion. Translated by R. F. C. Hull. Vol. 11. Collected works. New York: Pantheon Books, 1958.
[3] Jung, C. G. (1910). The association method. American Journal of Psychology, 21, 219--240.
[4] Jung, C. G. (1933). Modern man in search of a soul. New York: Harcourt, Brace & World.
[5] Jung, C. G. (1954a). The development of personality. Collected works (Vol.17), Bollingen Series XX, G. Adler, M. Fordham, & H. Read (Eds.) (R. F. C. Hull,Trans.). New York: Pantheon Books.
[6] Jung, C. G. (1954b). The practice of psychotherapy. Collected works (Vol.16), Bollingen Series XX, G. Adler, M. Fordham, & H. Read (Eds.) (R. F. C. Hull,Trans.). New York: Pantheon Books.
[7] Jung, C. G. (1958). Psychology and religion. Collected works (Vol. 11),Bollingen Series XX, G. Adler, M. Fordham, & H. Read (Eds.) (R. F. C. Hull,Trans.). New York: Pantheon Books.
[8] Jung, C. G. (1960a). The psychogenesis of mental disease. Collected Works (Vol. 3), Bollingen Series XX, G. Adler, M. Fordham, & H. Read (Eds.) (R. F. C.Hull, Trans.). New York: Pantheon Books.
[9] Jung, C. G. (1960b). The structure and dynamics of the psyche. Collected works (Vol. 8), Bollingen Series XX, G. Adler, M. Fordham, & H. Read (Eds.) (R.F. C. Hull, Trans.). New York: Pantheon Books.
[10] Jung, C. G. (1961). Freud and psychoanalysis. Collected works (Vol. 4). Bollingen Series XX, G. Adler, M. Fordham, & H. Read (Eds.) (R. F. C. Hull,Trans.). New York: Pantheon Books.
[11] Jung, C. G. (1963a). Memories, dreams, reflections. Recorded and edited by Aniela Jaffe'. New York: Pantheon Books.
[12] Jung, C. G. (1963b). Mysterium coniunctionis. Collected works (Vol. 4), Bollingen Series XX, G. Adler, M. Fordham, & H. Read (Eds.) (R. F. C. Hull,Trans.). New York: Pantheon Books.
[13] Jung, C. G. (1964a). Civilization in transition. Collected works (Eds.) (R.F. C. Hull, Trans.). New York: Pantheon Books.
[14] Jung, C. G. (1964b). Man and his symbols. New York: Doubleday.

[15] Jung, C. G. (1966). The spirit in man, art, and literature. Collected Works (Vol. 15), Bollingen Series XX, G. Adler, M. Fordham, & H. Read (Eds.) (R. F. C.Hull, Trans.). New York: Pantheon Books.
[16] Jung, C. G. (1967). Alchemical studies. Collected works (Vol. 13). Bollingen Series XX, G. Adler, M. Fordham, & H. Read (Eds.) (R. F. C. Hull,Trans.). New York: Pantheon Books.
[17] Jung, C. G. (1971a). Aion. In J. Campbell (Ed.), The portable Jung (R. F. C. Hull, Trans.) (pp. 139-162). New York: Penguin Books.
[18] Jung, C. G. (1971b). Personality types. In J. Campbell (Ed.), The portable Jung (R. F. C. Hull, Trans.) (pp. 178--272). New York: Penguin Books.
[19] Jung, C. G. (1971c). Psychological types. Princeton, NJ: Princeton University Press.
[20] Jung, C. G. (1971d). The stages in life. In J. Campbell (Ed.), The portable Jung (R. F. C. Hull, Trans.) (pp. 3--22). New York: Penguin Books.
[21] Jung, C. G. (1971e). The structure of the psyche. In J. Campbell (Ed.),The portable Jung (R. F. C. Hull, Trans.) (pp. 23-46). New York: Penguin Books.
[22] Jung, C. G. (1971f). The transcendent function. In J. Campbell (Ed.),The portable Jung (R. F. C. Hull, Trans.) (pp. 273-300). New York: Penguin Books.
[23] John S. Bell, "On the problem of hidden variables in quantum mechanics," Rev. Mod. Phys. 38, 447-452-1966.
[24] A. Einstein, B. Podolsky, and N. Rosen, "Can quantum-mechanical description of physical reality be considered complete?," Phys. Rev. 47,777-780-1935.
[25] John S. Bell, "On the Einstein Podolsky Rosen paradox," Physics 1,195–200 _1965_. Reprinted in John S. Bell, Speakable and Unspeakablein Quantum Mechanics _Cambridge U. P. Cambridge, 1987_.
[26] 91–94 _1982_; Alain Aspect, Jean Dalibard, and Gérard Roger, "Experimental test of Bell's inequalities using time-varying analyzers," ibid.
[27] Henry Pierce Stapp, "S-matrix interpretation of quantum theory," Phys.Rev. D 3, 1303–1320 1971.
[28] Henry P. Stapp, "Nonlocal character of quantum theory," Am. J. Phys.65, 300–304-1997.
[29] Aspect, A., Dalibard J., & Roger G. (1982). Phys. Rev. Lett. 49, p.1804.
[30] Herbert, N (1996). Quantum Reality (New York : Dutton).
[31] Friedman N (1994). Bridging Science and Spirit. Common elements in David Bohm's Physics Perennial Philosophy and Seth (MO : Living Lake books) Ch 1, pp.37-93
[32] Mukhopadhyay, A. K. (2002). Science for Consciousness. Five reasons for Failure and five Ways to make it a success. Frontier Perspectives, 11(1), pp.33-35.
[33] Walker, E. H. (1970). Math. Biosci. 7, p.131.
[34] Stapp, H. P. (1982). Found. Phys. 12, p.363.
[35] Goswami, A. (1989). Phys. Essays 2, p.385.
[36] Eccles, J. C. (1994). How the Self controls its Brain (New York : Springer-Verlag) Ch. 9, pp.145-166.
[37] Penrose, R., & Hameroff, S. (1994). Journal of Consciousness Studies 1(1), pp. 91-118
[38] Mukhopadhyay, A. K. (2000). The Millennium Bridge (New Delhi: Conscious Publications) Ch. 10, pp.170-195.
[39] Grinberg-Zylberbaum J., Delaflor M., Attie L., & Goswami A. (1994). Physics Essays 7 (4), pp.422-28.
[40] Long, M. F. (1976). The Secret Science behind Miracles: Unveiling the Huna Tradition of Ancient Polynesians (California : Devorss & company).
[41] Mukhopadhyay, A. K. (2000). The Millennium Bridge (New Delhi : Conscious Publications) Ch.7, pp.133-135.
[42] Hawking, S., & Penrose, R. (1997). the nature of space and time (Delhi : Oxford Univ. Press) Ch.4, pp.61-63.
[43] Hawking, S. (1988). A brief history of time (Toronto etc. : Bantam Books)
[44] Majumder, P. P. (2001). Ethnic populations of India as seen from evolutionary perspectives. J. Biosci, 26 (4) (suppl.), pp.533-545.

[45] Mukhopadhyay, A. K. (1995). Conquering the Brain (New Delhi : Conscious Publications) Ch.1, p.10 and Ch.18, p.122.

[46] Mukhopadhyay, A. K. (2000). The Millennium Bridge (New Delhi : Conscious Publications) Ch.6, p.110.

[47] Dennett, D. C. (1995). Darwin's Dangerous Idea: Evolution and Meaning of Life (New York : Simon & Schuster) p., 1,21,59, 82,335,394.

Made in the USA
Columbia, SC
04 September 2024